LANDMARK COLLECTOR'S LIBRARY

ALLDAYS & ONIONS

A BRIEF HISTORY

NORMAN PAINTING

LANDMARK COLLECTOR'S LIBRARY

ALLDAYS & ONIONS

• A BRIEF HISTORY •

Norman Painting

Landmark Publishing

Published by

Ashbourne Hall, Cokayne Ave
Ashbourne, Derbyshire DE6 1EJ England
Tel: (01335) 347349 Fax: (01335) 347303
e-mail: landmark@clara.net
web site: www.landmarkpublishing.co.uk

ISBN 1 84306 047 7

British Library Cataloguing in Publication Data: a catalogue
record for this book is available from the British Library.

Printed by CPI, Bookcraft, Midsomer Norton, Somerset
Design: Ashley Emery
Cover: James Allsopp

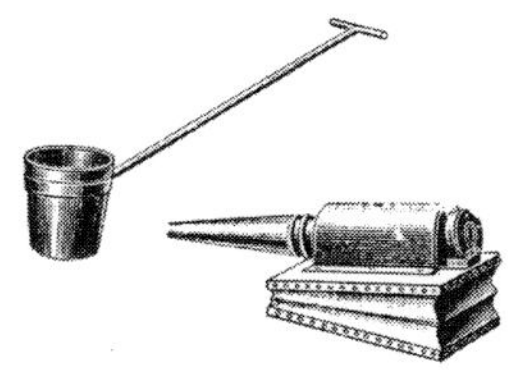
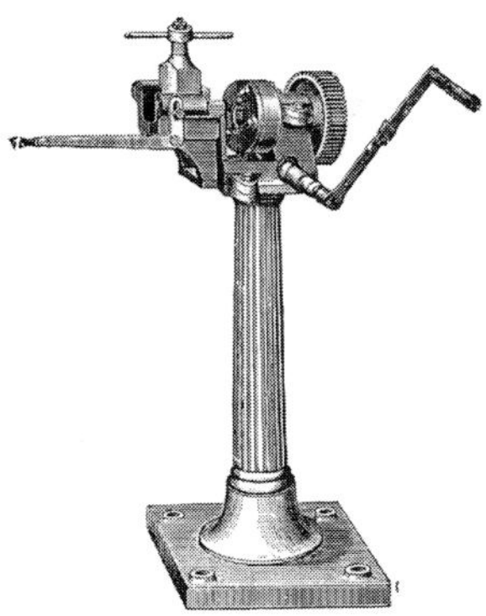
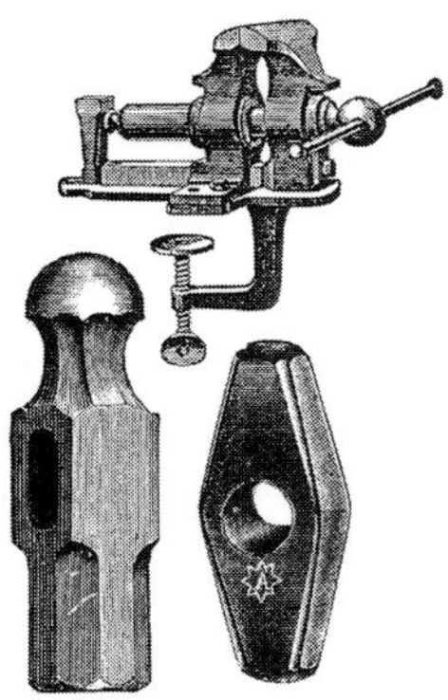

CONTENTS

ALLDAYS

EXTRACT FROM THE "DAILY NEWS," July 17th, 1920

"THE TRUTH UNLIMITED.

The war has left us a legacy of high prices ; but cars and motor-cycles in addition to their double cost, make a great inroad on people's incomes for petrol, oil, tax and wear and tear expenses. The bicycle, it is true, has shared in the general advance as regards first cost, but in motive power it costs no more to-day than in pre-war times ; while in the matter of repairs and upkeep, these items on a well-made bicycle are so small as practically to be not worth consideration.

The above is an extract from an introductory chapter to the bicycle catalogue of Messrs. Alldays & Onions, Ltd.—an excellent catalogue. The paragraph quoted is intended to suggest, among other things, that the Alldays is a "well-made Bicycle." So it is! I know, because the maker lent me one about two months ago. It is their "standard light roadster," with 26 in. wheels, 1⅜ in. open-side Dunlops, flat handlebar, rolling lever brakes, sloping top tube, ample mudgrards. Its price, I believe, is £15, and my spring balance says it weighs 31½ lbs. It is just exactly tha high-class production which you expect from a firm 270 years old, that had an ancient and honorabje reputation in general engineering work before they turned their attention and their skill to bicycles and motor-cycles"

CYCLES

Send for Catalogue—Free

ALLDAYS & ONIONS Ltd., BIRMINGHAM

ALLDAYS

FOR ALL TYPES OF PEDAL-DRIVEN CARRIERS

BACKED BY

HALF A CENTURY'S EXPERIENCE

IMPROVED TRANSPORT MODEL

Prices from £5.17.0

ALLDAYS MATCHLESS TRICYCLE.

ALL TRICYCLES HAVE INTERNAL EXPANDING BRAKES TO ALL WHEELS

Prices from £13.10.0

- **IMPROVED DESIGNS**
- **KEEN PRICES**
- **INCREASED PROFIT**

AS OUR MACHINES WILL NOT BE EXHIBITED AT OLYMPIA THIS YEAR, APPLY AT ONCE FOR THIS VALUABLE AGENCY.

ALLDAYS & ONIONS, Ltd.

GREAT WESTERN WORKS,

SMALL HEATH, BIRMINGHAM, 11.

London Office: 9, Victoria Street, S.W.1.

PREFACE

No written history can ever do justice to this company which was able to trace some of its ancestry back to 1625 and which, during more than 100 years, produced items as humble as a Bunsen Burner up to enormous fans which dwarfed the men building them. Between these two extremes they made a bewildering range of engineering products, mostly associated with what we in Birmingham call "The Metal Bashing Industries". Indeed, they made so many different products I found it difficult to decide what to leave out, but in the words of Voltaire, "The way to be a bore is to say everything."

With the remains of the Great Western Works just a few minutes walk from my home in Small Heath, Birmingham, I have often stood on the canal bridge in Golden Hillock Road looking at the works and trying to imagine the hive of activity it must have been in its heyday with smoke belching from the foundry chimneys, the noise of machinery and the vibration through one's feet as the drop forging hammers pounded white hot steel into shape.

However, this was but one of the factories owned by the company and through this brief history the progress of this great engineering company can be followed together with its foray into bicycle, motor cycle, commercial vehicle and motor car production, the latter bringing the company to its knees financially like so many other companies in the post First World War years.

Although it rose from the ashes and returned to the production of its general engineering range of products, the company was to remain totally within its original Great Western Works, eventually concentrating its engineering expertise on specialised industrial fans. In 1988 manufacturing was moved to Weston-Super-Mare, where Alldays had already been manufacturing fans since 1969.

This brief history is but a poor tribute to the two families who built up a company which could proudly boast the longest history of any engineering business in Britain and to the many hard-working, loyal and long-serving employees of this Birmingham company.

Norman Painting.
August 2002

ILLUSTRATIONS

These two advertisements appeared in "The Motor Cycle and Cycle Trader" journals dated September 24th 1920 and November 2nd 1934, respectively. The fourteen year span between these advertisements was to see the company go into receivership, two changes in the company name and a rapid decline in the cycle making side of the company's business.

The manufacture of tradesmen's cycles and tricycles had been quite profitable, but the mass production of motor cycles and, more particularly, the motor vehicle had made serious inroads into the retail delivery of goods. Partly as a result of this and the increasing availability of hire purchase to buyers and subsequent overheads caused by pursuing payment defaulters, the cycle business was wound up in 1935. The advert shown at the bottom of the page was, therefore, one of the last to be placed by the company before cycle production ceased.

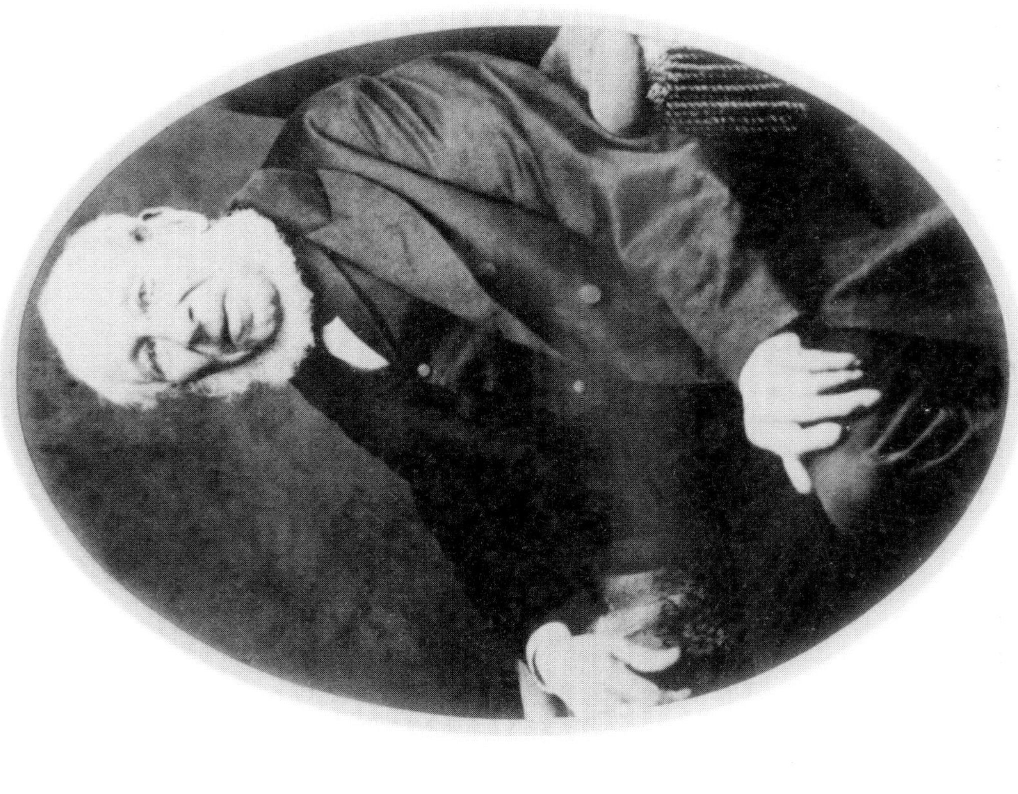

ALLDAYS & ONIONS PNEUMATIC ENGINEERING CO. LTD.

ACKNOWLEDGEMENTS

The writing of this history would have floundered at the first page but for the invaluable assistance given to the author by past employees of the firm, vehicle owners, interested parties answering the author's pleas for assistance and, above all, members of the Alldays family itself.

The author would like to express his sincere thanks to the following people, and hopes that no one is missed out from the list which follows.

Peter Thomas, George Corbett, John Bird, David E.A.Evans, David Wilkins, K. Halstead, Richard Moss, J. Taylor, K.F.Hillier, W. Arnott, Arthur Hunt, Jack Erskine, Wilf Gale, Johnny Thomas, Margaret Williams - Historical Researcher of Unilever UK Central Resources Limited, Derek Reid - BT Archives, Barry Collins - Museum of British Road Transport in Coventry, David Rossington (Australia), John Newland (Australia), Rob Knight (New Zealand), Graham Hallard (motor cycles), Jim Boulton - for a wealth of information and introductions, the staff of Local Studies in the Coventry Central Library, the staff of Local Studies Department and Archives Department of Birmingham Central Reference Library, the staff of Wallwork Ltd., Birmingham, who currently occupy the old Great Western Works, the staff of Alldays, Peacock & Co.Ltd., Weston-super-Mare, and finally, to Michael and Mary Allday for their hospitality and endless generosity in allowing the author access to their 'treasure trove' of family history, company records and technical publications.

PATENTS

Drawings from Patents are reproduced by kind permission of The Controller of Her Majesty's Stationery Office. Full copies of patents are available from the Patents Office, Sales Branch, Concept House, Cardiff Road, Newport, Gwent, NP9 1RH.

DEDICATION

This book is dedicated to the late Norman Hirst. Norman was a self made businessman who had served an apprenticeship at the B.S.A. works in Small Heath, and went on to establish his own business in Coventry called Coventry Boring and Metalling Co. Ltd. During the Second World War he had been a volunteer fireman in Birmingham and had helped to fight the fire at the Alldays factory which had been caused by bombing. The author received much encouragement and assistance from Norman whilst researching Alldays, for which he is eternally grateful.

The authors royalties from this book will be donated to the Royal Star and Garter Home for Ex-Service Men & Women

PHOTOGRAPHS

Two members of the Alldays family who were engaged in the engineering business in the 18th and 19th centuries.

The lower picture is of William Allday 1798 to 1882 and the top picture is of William Allday 1822 to 1888.

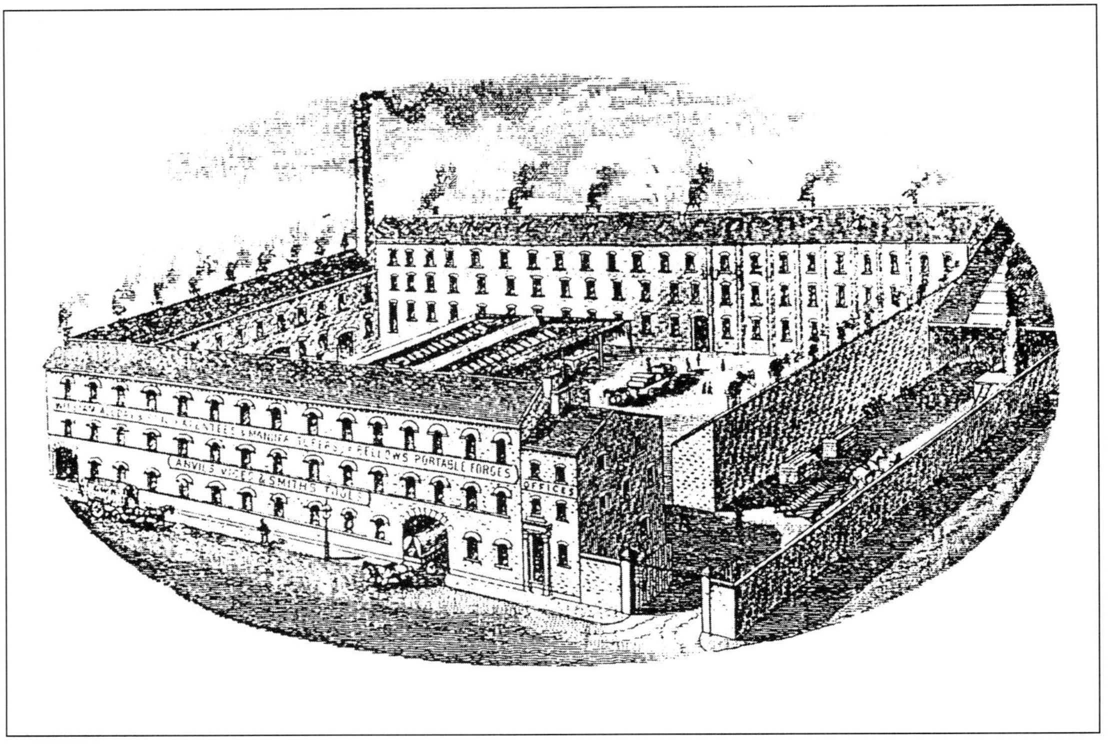
BELLOWS PORTABLE FORGES
ANVILS VICES & SMITHS TOOLS
OFFICES

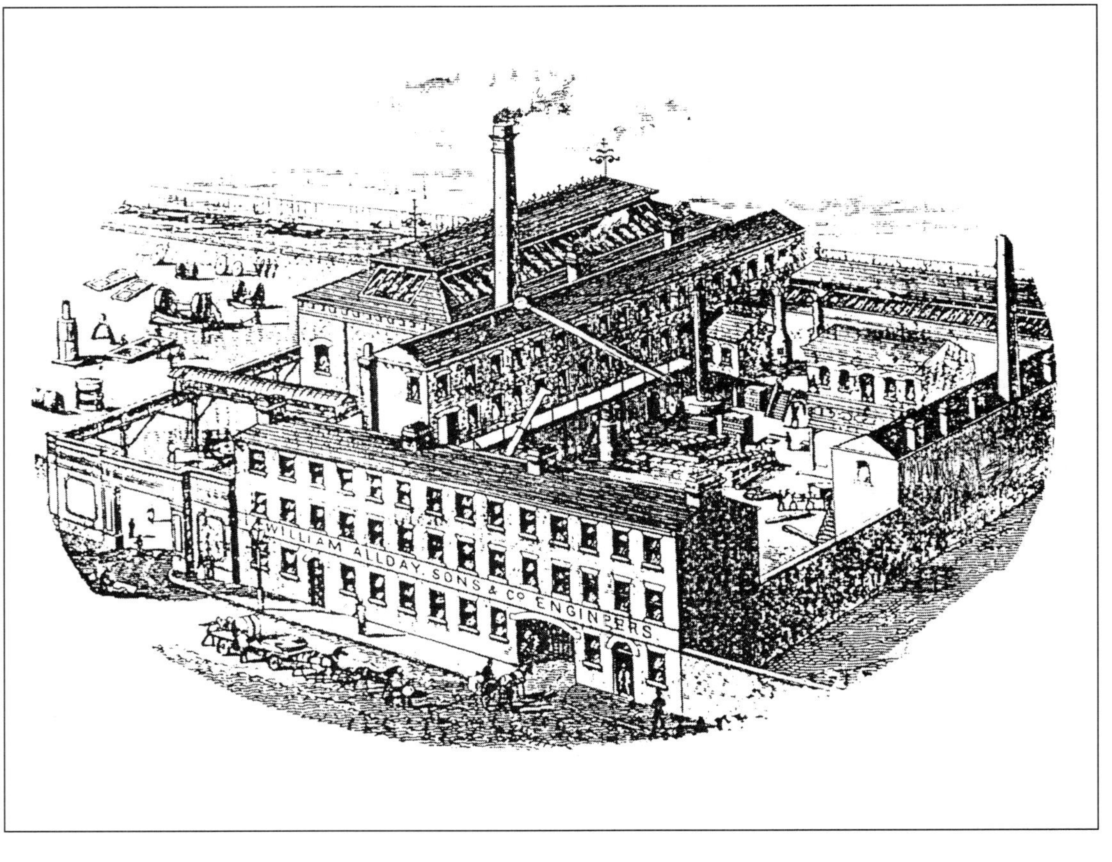
WILLIAM ALLDAY, SONS & Co ENGINEERS

THE ORIGIN OF THE COMPANY

Although the company claimed that it had been established since dates varying between 1650 and 1780 the union of these two Birmingham companies did not take place until 1885 when William Allday & Sons, of Branston Street, and J.C. Onions, of Bradford Street, merged to form the company of Alldays & Onions Ltd. However, a craftsman named Onions is believed to have been making bellows in a shed near Dudley Castle in 1625 and some twenty five years later the brothers Onions had established a business in Birmingham where, in 1770, a John Onions was known to have a bellows making business in Digbeth. By 1785 his brother Simon had joined the company which was then situated in premises behind the "White Swan" in Digbeth. For a short time the two brothers parted and Simon set up his own business at 12 Moseley Street. In 1823 they joined forces again and set up their business at 95 Bradford Street.

In 1854 the company presented a pair of ornamental library bellows to Napoleon, using material from a willow tree close to the grave of Bonapart, which was duly acknowledged by "The Emperor of the French" in May 1854. About 1884 J.C. Onions (John Collingwood Onions) opened a foundry in Charles Henry Street and also used premises in Oxford Street, Digbeth, for the storage of timber which was used in large quantities in the manufacture of bellows.

The company Alldays & Co. was founded in 1788 and by 1846 had become established makers of bellows and portable forges with premises at 126 Livery Street. After moving to premises in Constitution Hill in 1856 and then to premises in Great Hampton Street they finally settled down in much larger premises in Branston Street in 1875, under the name William Alldays & Son (Established 1780). By 1884 Alldays had set up their "Washington Foundry" in Commercial Street and also had the "Abberley Works" in Dudley which manufactured anvils, vices and smiths' tools, the company then trading under the name William Allday & Sons. (The sons being William the younger and Edmund)

An 1884 Alldays catalogue lists anvils from £1-3s-0d to £1-6s-0d per cwt (wholesale prices as sold to the trade) to which could be added an anvil stand at 9s-0d per cwt. Vices were priced at £1-10s-0d per cwt and a set of farriers' tools could be bought for £1-12s-6d. Ordinary smiths' bellows, but of the highest quality, ranged from a size 16" at £1-8s-0d per pair up to a size 50" at £32 per pair. A Number 9 double blast, hand operated, circular bellows in an iron frame for a smith's hearth cost £5-15s-0d for a size 18" up to £26-0s-0d for a size 40". The company had received several awards for its products including two gold medals for bellows and portable forges and iron hearths and grindstones and a silver medal for anvils, vices and smiths' tools at the 1880 International Exhibition held in Sydney, Australia, and also three highest awards at Melbourne, Australia in the following year for bellows, portable forges and grindstone troughs. Medals had also been awarded to the company in London in 1851 and Birmingham in 1879.

These competing companies were both heavily involved in supplying tools and equipment for what is commonly called 'the

ILLUSTRATIONS

These two illustrations taken from a 1884 William Allday & Co. catalogue show the Head Office and Works in Branston Street, in the Hockley area of Birmingham (top) and the "Washington Foundry" in Commercial Street (bottom). Both works were employed making engineering equipment and machines, but the Commercial Street works contained the main Iron Foundry. Significantly, the Commercial Street works backed onto canal wharves ensuring supplies of raw materials could be transported, and unloaded, direct into the premises.

As well as making the company's own wide range of products they were also engaged as Contractors in engineering and iron foundry work.

Surprisingly, the Branston Street works, which was a freehold property, and the Commercial Street premises, which was a leasehold property, were still owned by Alldays in 1911 when their value on the company balance sheet for that year was shown as £4,962-14s-1d.

THE ADMIRALTY

NEWLY-DESIGNED IRON-ENCASED PORTABLE FORGE,
STRONGLY RECOMMENDED FOR GENERAL OUT-DOOR AND RAILWAY WORK
AND EXPORTATION.

N°A 40.

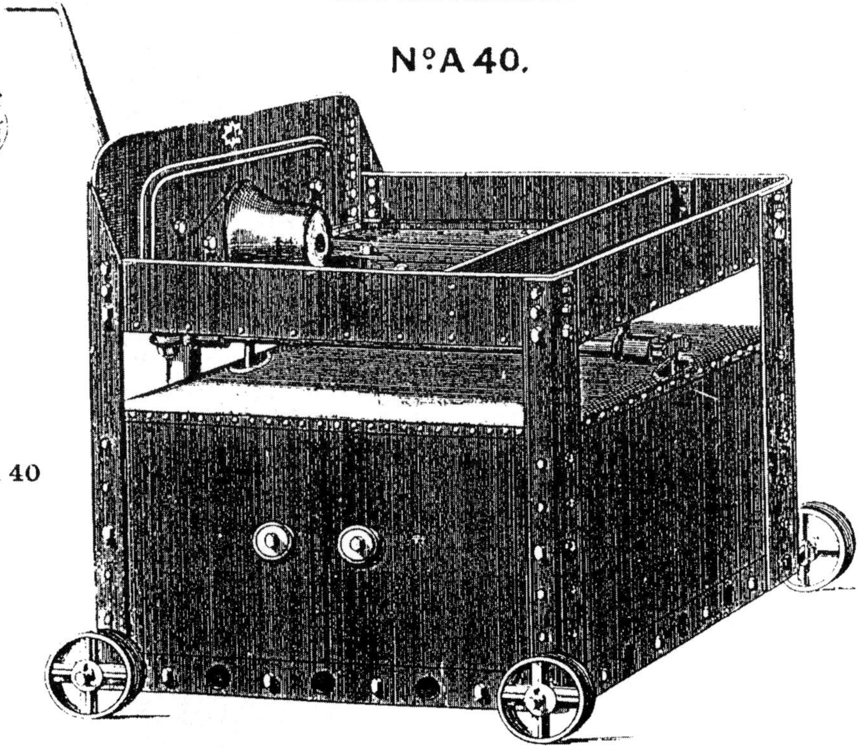

No. A 40

CONSTRUCTED BY REQUEST OF THE ADMIRALTY FOR USE IN THE ROYAL NAVY, MAY 15th, 1877, AND ORDERED FOR SUPPLY 1877 AND 1878; AND AGAIN ORDERED MARCH 14th, 1878, FOR SUPPLY, 1878 AND 1879; AND AGAIN ORDERED MARCH 3rd, 1880, FOR SUPPLY, 1880 AND 1881.

N°25.

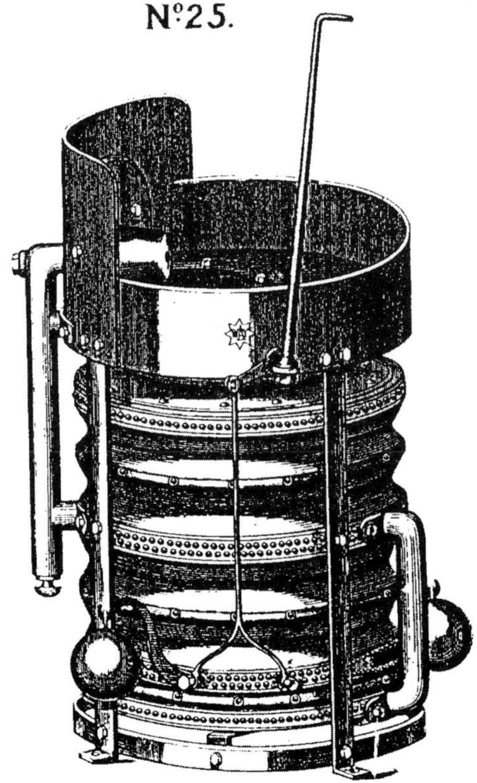

metal bashing industry' which has played a major role in the establishment of various industries in Birmingham and the Black Country and one cannot underestimate the value the humble bellows played in industry at that time. In 1866 the Onions company alone, which employed about 100 men by this time, was producing 30 forge bellows, 100 smiths bellows and over 3,000 house bellows per week, the latter of course being essential to householders who were reliant on coal fires in their homes for both heating and cooking. Some of this work was undertaken as 'outwork' carried out by men working, quite literally, from sheds in their back gardens, a practice still being carried out into the 1950s. This company also introduced a fan forge in which the blast of air was produced by turning a handle which rotated a fan and they also introduced steam driven circular saws and band saws.

Following the merger the company operated from the two premises, but it soon became apparent that their rapid growth, as a major supplier to the engineering industry in this country and abroad, would necessitate a move to larger premises and, in 1888, a factory was bought in Sydenham Road, Small Heath (just off Anderton Road) which had previously been occupied by the Midland Joinery Company. Sydenham Road at this time was a cul-de-sac which had not then been connected to the busy Golden Hillock Road. However, the site offered excellent prospects as the Grand Union Canal ran through the back of the site bringing in coal, sand and other raw materials direct to the factory wharf and the Great Western Railway ran virtually parallel with the canal, about 200 yards away. The only other occupant of Sydenham Road at that time was Thomas Francis who made iron nails and to the South and South West of the factory there were fields, together with Green Stile Farm which had a herd of milking cows which could be seen from some of the office windows.

In 1889 the company's name was changed to Alldays & Onions Pneumatic Engineering Company Limited, and the Sydenham Road factory was grandly named the Great Western Works.

ILLUSTRATIONS

LEFT: "The Admiralty" portable forge as illustrated in a 1884 catalogue. This was described as, "specially constructed for, and highly approved by, the Admiralty for use on board the Iron clads. It is made entirely of Wrought Iron, the bellows are of an improved pattern, producing an exceedingly strong and steady blast, and being iron encased, they are protected from injury; also rendered most adaptable for exposure in any climate; this forge is extensively used by Railway Constructors and is suitable for all kinds of outdoor work." That seems to say it all! This forge was made in three sizes, No.A40, A41 and A42. The circular forge on the right was described as "very superior, adapted to the use of Ship building, Dockyards, Steam vessels, etc." This forge stood 2 feet 10 inches high, was 2 feet in diameter and was fitted with 22 inch patent double-blast circular bellows.

BELOW: A view of the "Abberley Works" in Dudley where anvils, vices and smiths' tools were made.

ALLDAYS "CLIMAX" ROOT'S BLOWER CAST IRON PORTABLE FORGES

Nos. 10 and 11.

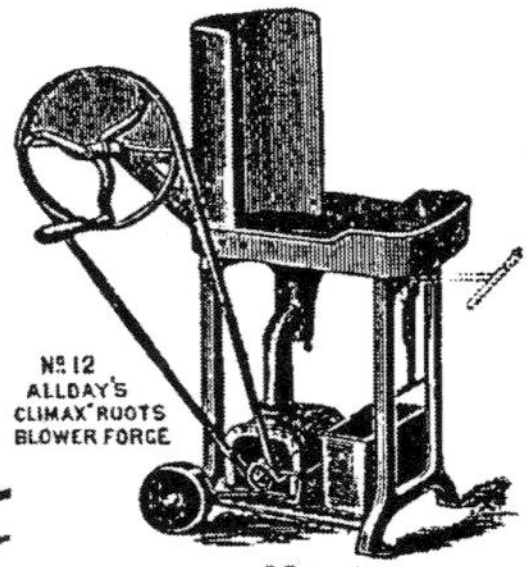

No. 12.

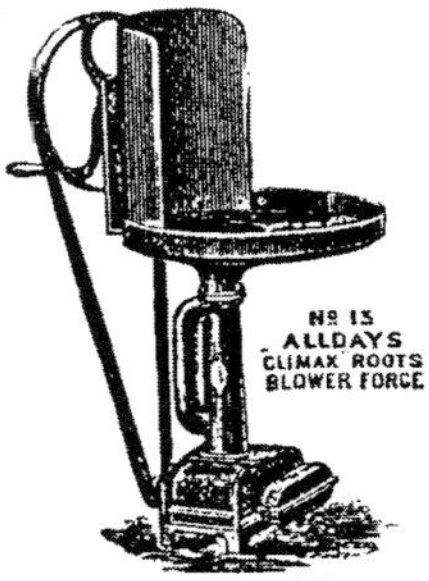

No. 13.

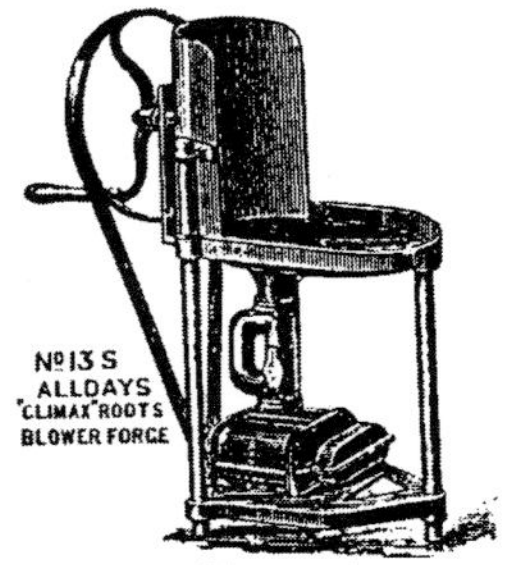

No. 13 S.

No.	Size of Hearth Plate.	Price. £	s.	d.	Travelling Wheels extra.	Approximate Weight. Cwt.
10	3' 4" × 2' 2"	12	5	0	15/-	6
11	2' 6" × 1' 9"	9	5	0	15/-	4
12	1' 8" × 1' 2"	6	5	0		2½
13	1' 9" × 1' 3"	4	15	0		1¾
13 S	1' 9" × 1' 3"	5	15	0		2

No. 12 is always mounted on wheels, which are included in the price. No. 13 does not require wheels.

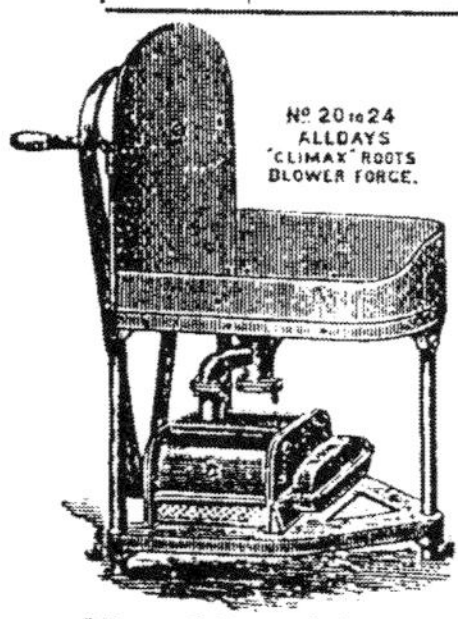

Nos. 20 to 24.

ALLDAYS "CLIMAX"
ROOT'S BLOWER MALLEABLE IRON PORTABLE FORGES

No.	Size of Hearth Plate.	Price. £	s.	d.	Travelling Wheels extra.	Approximate Weight. Cwt.
20	2' 2" × 1' 9"	9	5	0	12/-	2½
21	2' 2" × 1' 9"	7	2	6	12/-	1¾
22	1' 8" × 1' 6"	6	10	0	12/-	1½
23	1' 8" × 1' 6"	5	12	6	12/-	1¼
24	1' 4" × 1' 2"	5	0	0	Not required.	1
25	2' 2" × 1' 9"	6	12	6	12/-	1½
26	1' 6" × 1' 4"	5	10	0	12/-	1¼

Nos. 25 & 26, similar to drawing No. 24, with the exception that they have bac! blast instead of bottom. Spare Tue Irons for Nos. 25 & 26, 2/6 each; all th other numbers 6d. each.

No.		Size of Hearth Plate.	Approximate Weight. Cwt.	qrs.	lbs.	Equal to Bellows.	Capable of Heating Iron Bars.	Net Price, Delivered at our Works. £	s.	d.
As per Drawing No. 3	1	20 in. diam.	1	2	0	24 in. diam.	1¼ in. square.	6	0	0
	2	22 ,, ,,	1	3	0	28 ,, ,,	1½ ,, ,,	8	0	0
	3	24 ,, ,,	2	0	21	36 ,, ,,	2 ,, ,,	10	0	0
	4	26 ,, ,,	2	2	0	40 ,, ,,	3 ,, ,,	12	0	0

Having received many requests from Home and Foreign Government Officials, Ship Owne and Builders, and large Contractors, for Portable Forges other than the soft fan blast, with entire absence of wood, leather, or any parts liable to get out of order, decay, or destructio but still possessing a positive or force blast, and knowing that **Allday's Root's Pater Improved Pressure Blowers** possess the only means that will in every way meet with the above requirements, we have, after considerable experimenting, pleasure in introducing the **"Climax"** Portable Forges, which for compactness, great strength, durability, simplicit **ease of working, and great volume of pressure or positive blast, cannot** be **equalled.** It is adapted for fitting and erecting shops or railway extensions at home, **exposure** on board ship, or in use in either **Tropical** or **Arctic** climates, and we are su upon trial, our numerous clients will agree with us that we have rightly named them t **Climax of all Portable Forges.**

Branston Street Works, Birmingham.

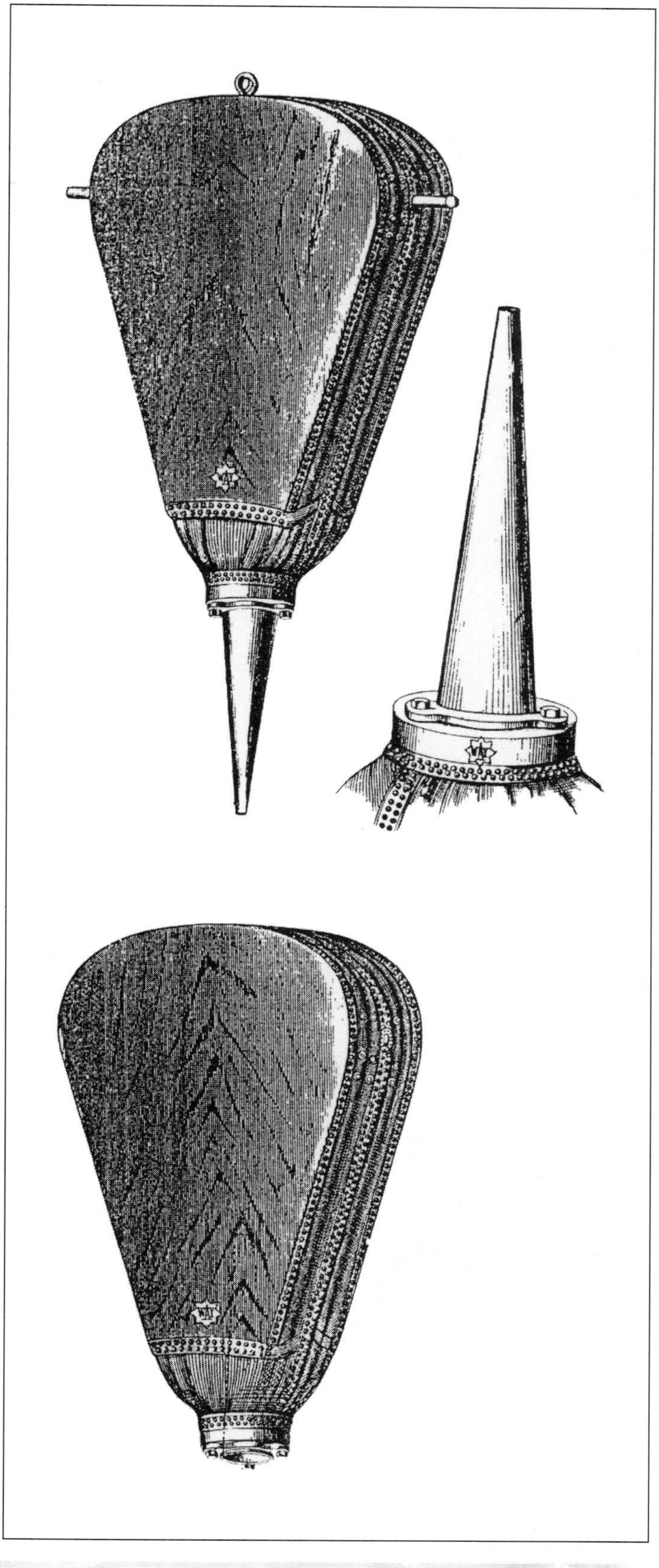

ILLUSTRATIONS

LEFT: Some of the "Climax" range of portable forges being offered by the company in 1884 with the blast for the fire being produced by hand operated Root's blowers. Prices ranged from £4-15s-0d to £12-5s-0d and the largest forge in this range could heat a 3" square steel bar to a temperature high enough for it to be worked on an anvil.

RIGHT: This illustration shows the patented "Screw Pipe Smiths Bellows" in which the pipe could be unscrewed, reversed and fastened safely inside the bellows itself for storage and/or transport. This feature was particularly advantageous for shipping abroad when the packing case size could be reduced by almost one third, saving shipping costs and thereby gaining a little advantage pricewise in a very competitive market. These bellows were made in sizes from 16 inches to 50 inches long and with prices from £1-8s-0d to £32 per pair. Bellows were to remain a principle product of the company even into the 1930s and in 1916 the company employed as many as 250 workmen making bellows alone.

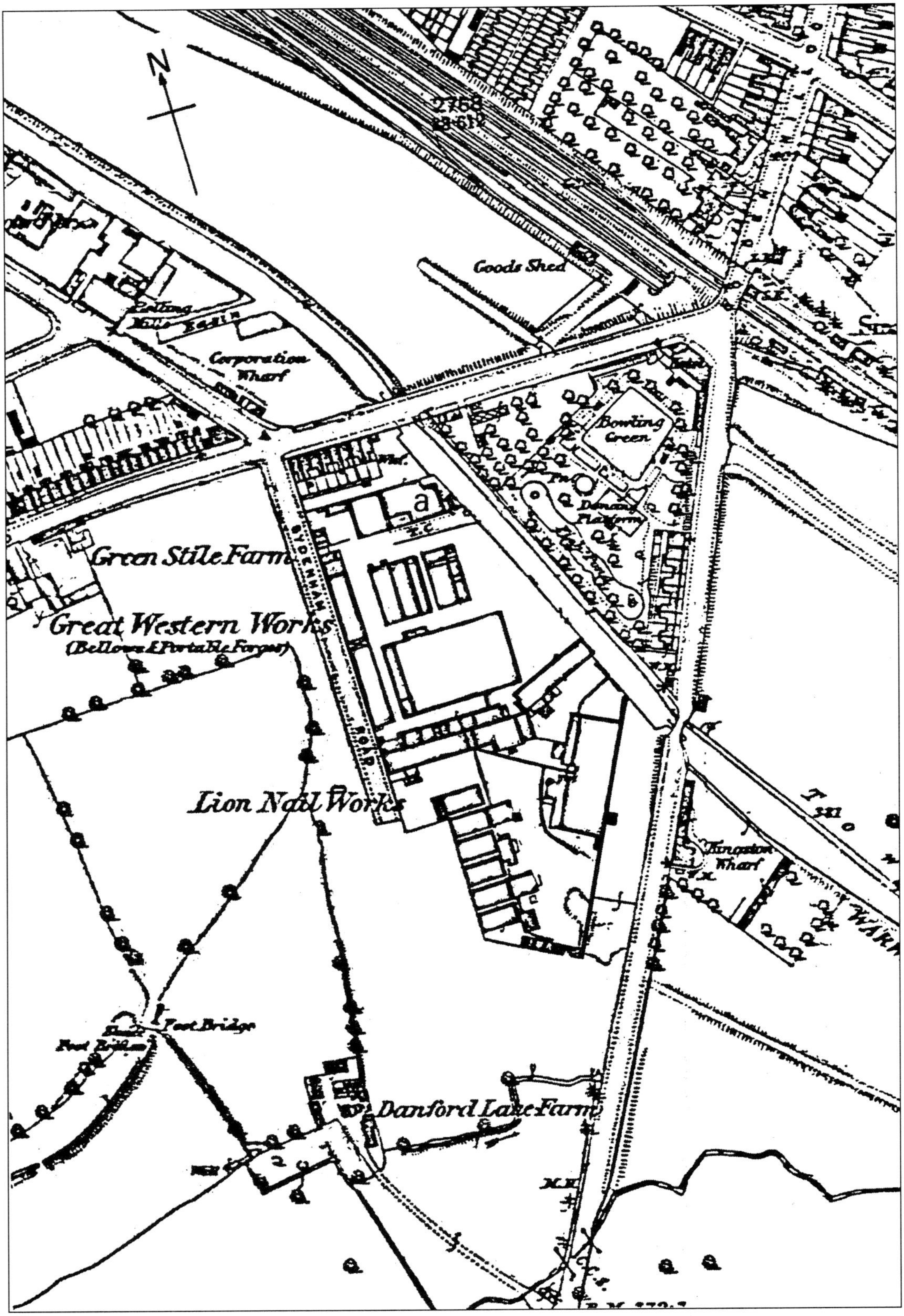
N
Goods Shed
Corporation Wharf
Bowling Green
Green Stile Farm
Great Western Works
(Bellows & Portable Forges)
SYDENHAM ROAD
Lion Nail Works
Foot Bridge
Danford Lane Farm

GREAT WESTERN WORKS

Within a few months of moving into the Sydenham Road factory the company started to extend the facilities, the first of which in October 1888 brought about an extension to the Foundry. This single storey building measured 66 feet by 58 feet and, interestingly, used an old roof removed from a building in the Branston Street works. The architect employed for this work was George Kenwrick of 83 Colmore Row, Birmingham.

The second building to be erected in 1888 was an extension to the Core Shop which comprised two Core Stores and a Core Shop complete with a core stove. The Core Shop itself was 25 feet by 26 feet 9 inches and this single storey building was 19 feet high to the eaves and 30 feet high to the ridge. The drawing for this building states that the foundations were to be of "Hardest Darley Dale Stone".

Apparently finding difficulties in recruiting workers for the factory the company embarked on an ambitious scheme in June 1894 to provide company housing hard by the works. The architect Ewen Harper of 57 Colmore Row was employed to draw up plans for 39 houses which were built on land at the south corner of Sydenham Road and Anderton Road. The houses were built with 9 properties along Anderton Road, two rows containing 11 houses to the rear of and parallel to the first 9 and the final 8 houses to the rear of the second row of 11 properties and positioned well back from Sydenham Road. This was a bold and expensive plan to undertake, but one can only applaud the actions of the company to provide decent housing for its workforce.

The factory was now fully employed in the manufacture of industrial and household bellows, water turbines, centrifugal fans, positive pressure blowers, pneumatic hammers for the forging industry, cupolas and furnaces for castings and a wide variety of tools and workshop equipment for more general engineering.

A patent was taken out by William and Edmund Allday in 1893 for improvements to engineers' bench vices and was followed in 1894 by a patent relating to improvements to bellows for portable forges. There was obviously an intense driving force within the company, not only to make the best product, but to improve it whenever possible to stay ahead of the competition. Two patents were taken out in 1894 for ambulances of a type we would find difficult to comprehend today, but were, nevertheless, considered to be of great value then as four of the cycle type ambulances were presented to Birmingham City by the Birmingham Hospital Saturday Fund. The company was already taking an active interest in the popular mode of transport of the day, the bicycle, and decided to build a second factory close by solely for the production of cycles. Work started on the new factory in 1895 just a short distance from the Great Western Works in Fallows Road which, itself, was in the early days of construction. To complete their housing scheme two more 3 bedroom houses were built in 1896 together with a Mess Room which became a social club for workers and their families in out-of-work hours. The Mess Room was named Great Western Hall and was located opposite the works in Sydenham Road. This two storey building measured 71 feet by 32 feet and contained a shop 16 feet by 15 feet between the two ground floor entrances, one of which led directly into the ground floor room and the other accessed the upper floor via a stairway. Designed by Ewen Harper this imposing building was 24

ILLUSTRATION

This map shows part of the Small Heath district about 1890. Sydenham Road at this time ended abruptly in the fields of Danford Lake Farm, but would soon be extended to join up with the Golden Hillock Road which runs almost vertically to the right hand side of the map.

The precise extent of the Alldays & Onions factory at this time can be seen as the buildings from the start of the name Sydenham Road to its end, the site having its farthest boundary at the canal which runs at an angle of about 45° across the map.

The building identified with the letter 'a' housed the Foundry and Core Shop. Just below this building and marked 'TC' was a travelling crane mounted on girders 20 feet apart and extending to the canal side allowing barges to be unloaded direct into the factory grounds.

Fallows Road where the company would later build its cycle factory did not exist, but its approximate position would be in the area occupied by buildings making up Green Stile Farm close to Anderton Road which runs at the top of Sydenham Road.

The company had retained a small works at 14 Newton Street in the city centre, but this was closed about 1893 and all manufacturing was then conentrated in the Small Heath factory.

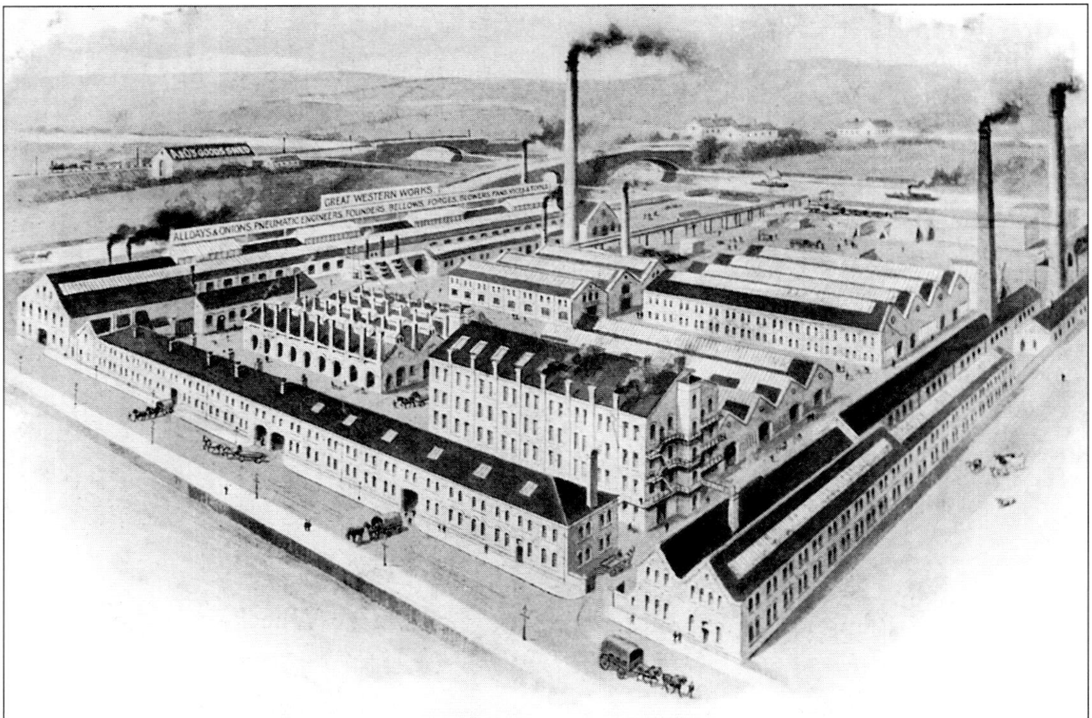

Great Western Works.

Alldays Parallel Vices of Wrought Iron and Steel.

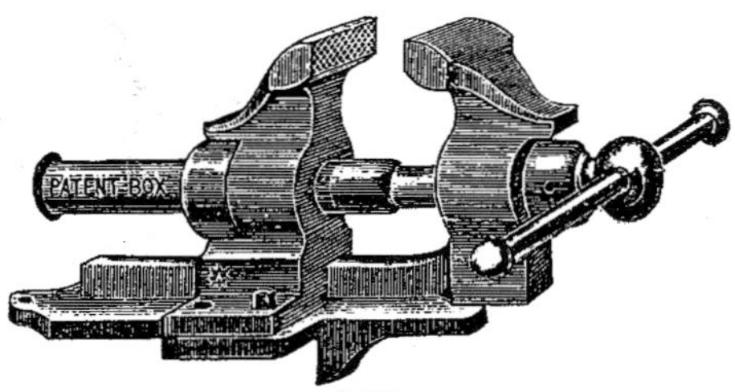

4127

3	3½	4	4½	5	5½	6	7	8 inch.
27/-	29/6	32/-	39/6	47/-	54/6	62/-	80/-	96/- ea.

4127 A

With front jaw to swivel in socket for taper grip.

3	3½	4	4½	5	5½	6	7	8	inch.
39/-	41/6	44/-	52/6	62/-	69/6	77/-			each.

T Bar Parallel Vice.

As supplied to H.M. Dockyards.

4123

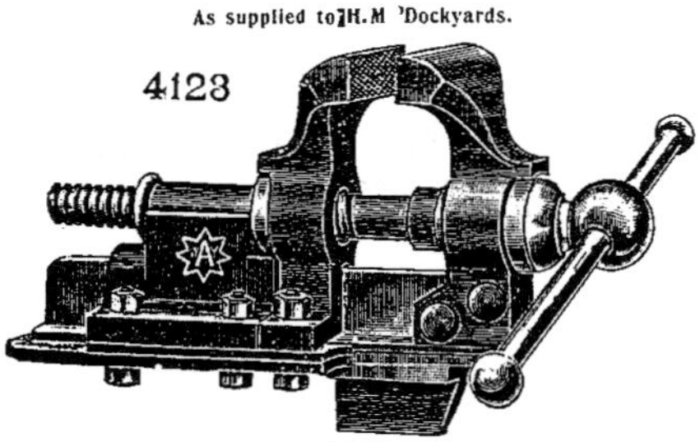

4123

3	3½	4	4½	5	6	7	8 inch.
34/-	36/-	38/-	40/-	43/-	63/-	85/-	100/- each.

Plumbers' and Amateurs' Vice.

4129

This 4in. Vice will allow a mechanic to do work in it with a 14in. bastard file without giving way.

3	3½	4 inch.
22/-	24/-	26/- each.
8	10	12 lbs.

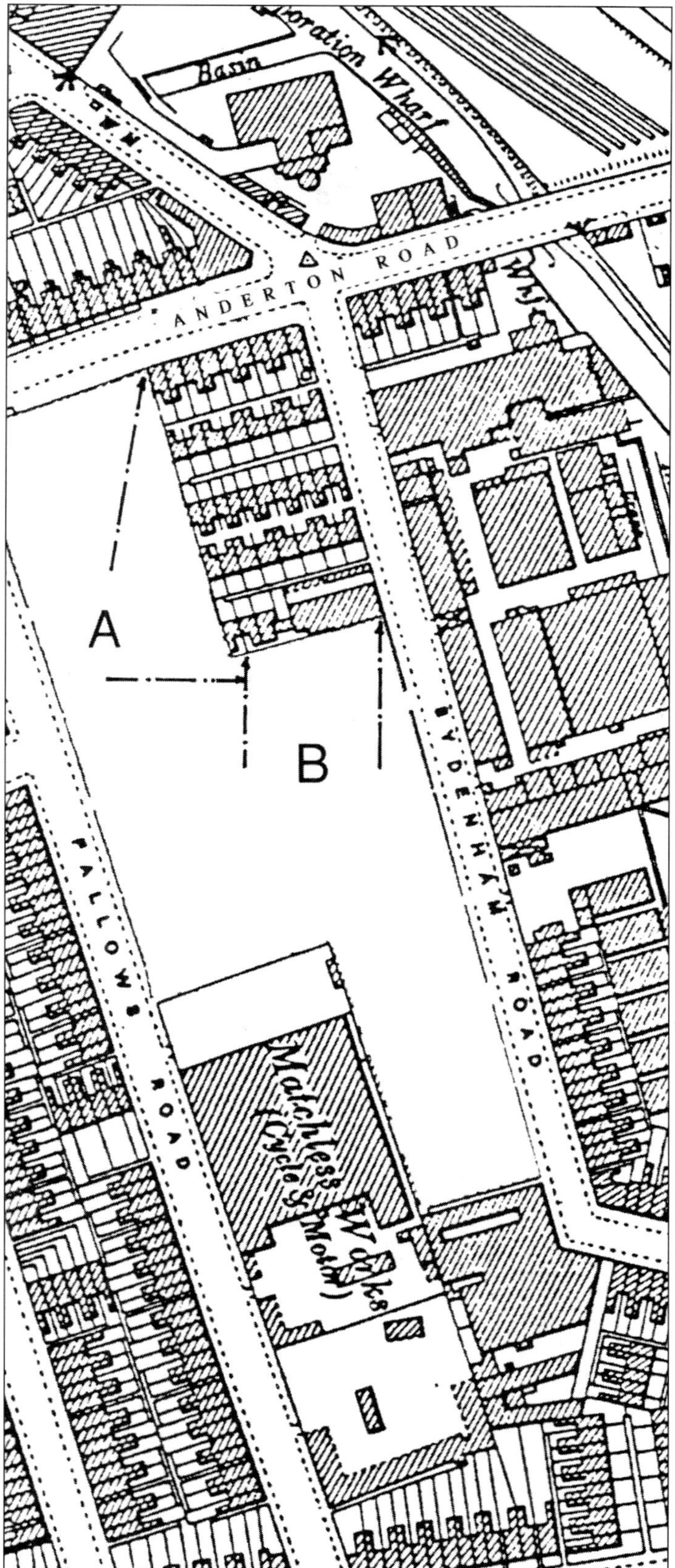

ILLUSTRATIONS

TOP-LEFT: Artist's impression of the Great Western Works, Sydenham Road in 1900. Although it gives some idea of how the works looked at that time it contains several errors, such as the railway line in the background crossing the canal, which actually runs parallel to it, and the long row of buildings to the right of the picture which, in reality, cut in at an angle towards the centre of the site. However, this picture, with slight modifications, such as changing the horse drawn carts for motor vehicles, was used in company literature until about 1920.

BOTTOM-LEFT: Just four examples of vices made by the company in 1905, the prices here ranging from £1-2s-0d up to £5. The 1905 catalogue contained about 28 different kinds of vices and this range would be increased quite substantially over the years.

RIGHT: This drawing is based upon a 1906 map of Small Heath and is used to illustrate the position of houses built by the company for its workforce. Some of the buildings in Anderton Road, Fallows Road and Sydenham Road have been removed for clarity.

Area 'A' shows that the 1894 proposal to build 39 houses at the corner of Sydenham Road and Anderton Road was modified slightly and in fact only 37 houses were built. Area 'B' shows where two further houses and the Mess Room cum Social Club was built in 1896.

To the bottom of the map can be seen the Matchless Works also built in 1896 in Fallows Road.

From these two maps can be judged just how rapidly Birmingham was being developed with housing and factories pushing their way into large areas which had been farmland and the Alldays workers could no longer watch cows being milked at local farms from their factory which had originally been surrounded by fields!

Alldays "Climax" Wrought-iron Smiths' Hearths.

Specially designed for the Admiralty, and with which we fitted up complete Her Majesty's Dockyard, Malta.

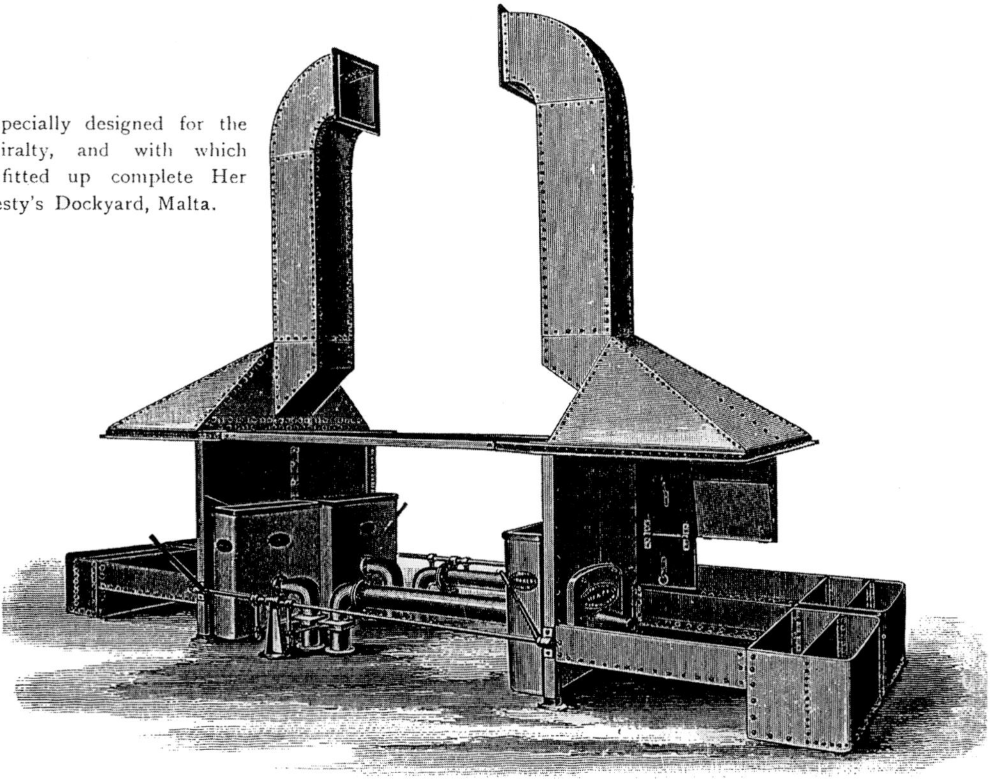

Alldays "Caledonian" Smiths' Hearths,

With and without Hoods.

Drilling Machines

Pattern No. 7A.

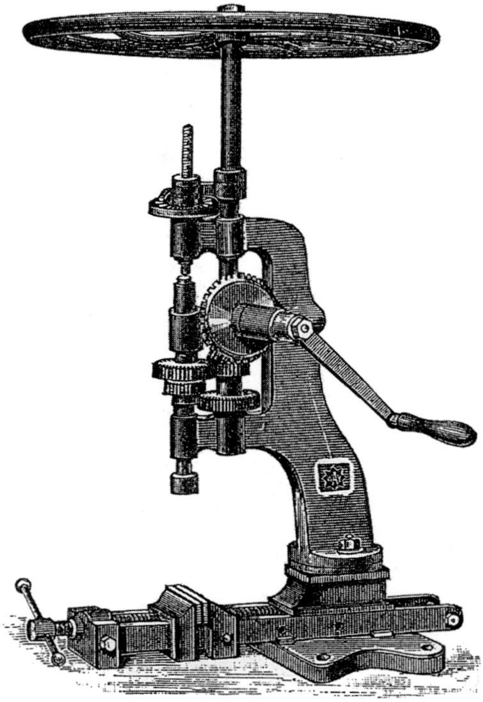

Pattern No. 12A.

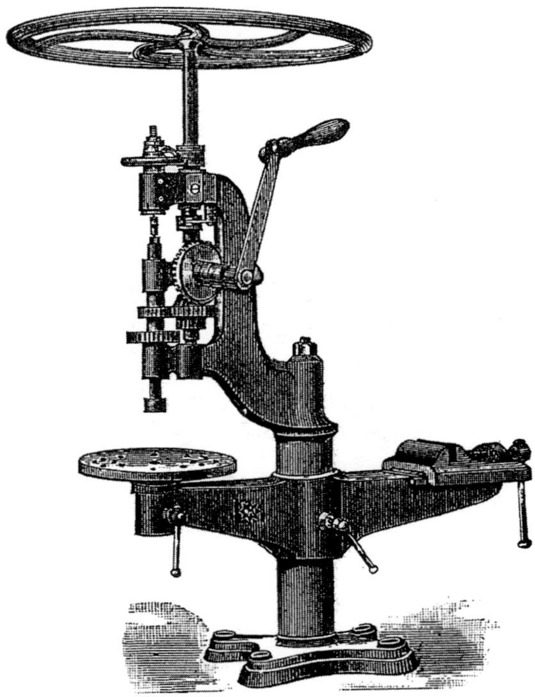

ILLUSTRATIONS

TOP LEFT: A coupled arrangement of two "Climax" double type smiths' hearths of a type supplied to large workshops where the blast air would have been supplied either by a Root's blower, or by a blower fan, the air being piped underground to the hearths and controlled by valves and levers shown to the rear of each hearth. The chimneys would probably have been coupled together into a common pipe to exhaust above the factory roof and all the necessary ducting was available from the company. These particular hearths were available in sizes from 6ft x 3ft up to 7ft x 5ft and the troughs at the front of each hearth were for coal and water.

BOTTOM LEFT: The "Caledonian" single hearths were available in sizes from 3ft x 3ft up to 4ft x 4ft and as double hearths too in sizes up to 6ft x 4ft and with the hood in place stood 6ft high.

RIGHT: These are just two examples of drilling machines available in 1900. Pattern Number 7a was of cast iron construction, had 2 speeds, complete with sliding parallel vice, could drill up to 1.25" diameter holes and cost £8 for a manually operated machine and an extra 15s-0d fitted with fast and loose pulleys for belt drive. Pattern Number 12a was of similar construction, but featured a circular drilling table as well as a parallel vice, both mounted on a cast arm which could be turned through 360° and raised or lowered and would accept work up to 26" in diameter. This machine cost £10-10s-0d in manual form and £11-10s-0d in belt driven form.

No. 2. Gearea

No. 26

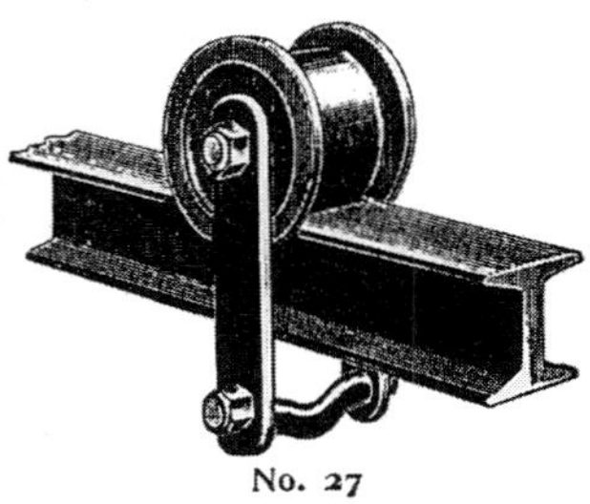

No. 27

No. 21.

Double Purchase, Cast-Iron Sides, Ordinary Lever Brake.

No 24.

Double| Purchase, Cast-Iron Sides, Screw Brake, Large Diameter Barrel for Wire Rope.

feet high to the eaves and 36 feet high to the ridge. As well as social gatherings and providing works mess facilities the new hall was also used as a Sunday School for the Mission of Christ Church in Dolobran Road close to the factory.

About 1899 the company acquired a business known as Eades Hoist Co. Ltd., who manufactured blocks and lifting gear and William Thomas Eades, who lived at 142 Walford Road, Sparkbrook, Birmingham, and was described simply as an engineer with the company, took out two patents in the same year dealing with a two speed warehouse hoist and crab winches. These new products were manufactured in the Great Western Works and were still marketed under the Eades Hoist company name until the First World War, after which they were marketed under the Alldays name. The largest hand operated double purchase hoisting crab with a 7" diameter barrel was capable of lifting 3 tons 12 cwts direct from the barrel and up to 18 tons with two and three sheave blocks and cost £17-12s-0d. Snatch blocks, Leading blocks, Cargo blocks, rigging and coupling screws and coal and rubbish gins were also among catalogue items for lifting tackle. "Elephant" brand warehouse and wharf trucks, which looked like wheeled modern day pallets, were also made for moving heavy loads in capacities from model number 4 for light loads at 5s-0d each up to model number 1 which measured 24"x 36", could carry 2½ tons and cost £2-2s-0d each.

Products from this works also included weights and measures in both English and metric values and foghorns made under the trade names of "Triton" and "Eclipse" which had been officially adopted by the British Admiralty. A letter from the Director of Navy Contracts was quoted in a 1900 sales catalogue thus:- "Gentlemen, I have to inform you that your tender of the 1st inst., has been accepted for the delivery of the under-mentioned articles at the yards stated. Bellows complete with Fog Horns, Key's Patent, Number 2 size. Deptford 30, Chatham 24, Devonport 60. - Director of Navy Contracts."

The humble household bellows were still being produced in large numbers from the basic Bastard Pattern No.180 at £1 per dozen for an 8 inch size up to £5-10s-0d per dozen for 11 inch Best Improved Long or Bristol Neats Leather type. Fancy household bellows were also being made with elaborate carved wood and ornate brass and japanned tops. Prospectors bellows for gold miners cost from 15 shillings per pair up to £2-9s-6d for a pair of 24 inch double blast type. Neat little bee smoking bellows were also made alongside enormous Smiths bellows up to 4 feet 2 inches long which cost £32 per pair, all made from best quality British materials and, it was said, that the biggest users of elm in Britain, were bellows and coffin makers!

Forges were made in a bewildering number of types and sizes, both static and portable, the latter models ranging from an 18 inch diameter model up to the largest 34 inches long by 30 inches wide and fitted with a 28 inch diameter double blast bellows. Forges with air supplied by a fan were also produced with names such as the "Branston", "Uneclipsable", "Empire", "Midget" and "Eureka", the latter model, which folded up into a box 3 feet long by 2 feet wide by 1 foot deep, having been specially designed for the Navy. A full range of blacksmiths' tools was made and sold to the trade by the lb rather than per item! Anvils were made for farriers, coachsmiths, chain makers, saw makers, edge tool makers and boiler makers, together with others with names such as London pattern, Cricket,

ILLUSTRATIONS

A large range of hoists, cranes and general lifting tackle was manufactured by the company, much of it under the name of Eades Hoist Co. Ltd. [The Eades Hoist Company, under the managership of William Thomas Eades, had been located at 40 Floodgate Street, Birmingham.] This page illustrates a few examples of the products being offered in 1905. In the top left corner is a 5 cwts Safety Power Hoist driven by a belt which featured a reverse motion device consisting of two cone friction pulleys operated by the chains A and B.

In the top right corner is a Travelling Worm Block fitted with a self-sustaining brake and made in a range from 10 cwts to 10 tons. Complete with chains for a 10 feet lift these blocks cost £7-15s for a 10 cwts model up to £40 for a 10 ton model.

The centre row of pictures shows, from the left, a No. 26 Overhead Travelling Runner to carry loads of 1 to 1.5 tons and cost £1 and £1-5s respectively, a No.27 Overhead Travelling Runner for loads of 1 to 4 tons, costing from £1-5s to £3, and the final illustration is a Light Crab used for lowering and raising arc lamps for lighting and cost 10s-6d.

The last two illustrations at the bottom of this page show two examples of Crab Winches. No.21 made in sizes from 16 cwts up to 3 tons 12 cwts cost £4 to £13-4s. The No. 24 model catered for loads of 8 cwts to 2 tons direct from the barrel, or 2 tons to 10 tons through sheave blocks and cost £4-10s up to £19 for the largest model.

Alldays Perfecta Patent Pneumatic Power Hammer.

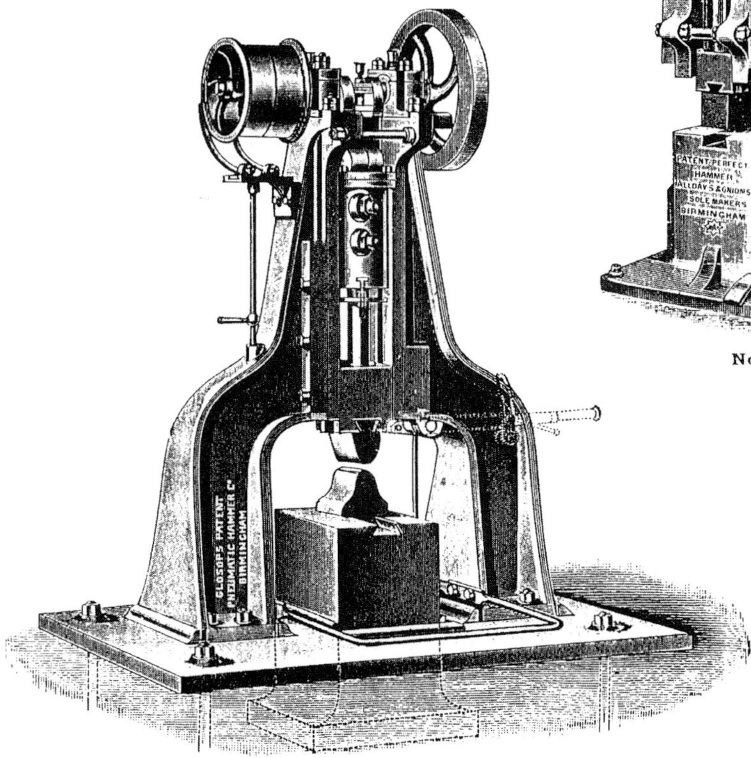

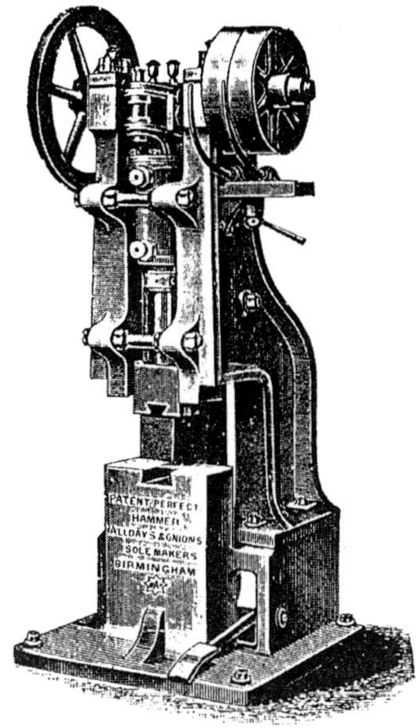

No. 3.

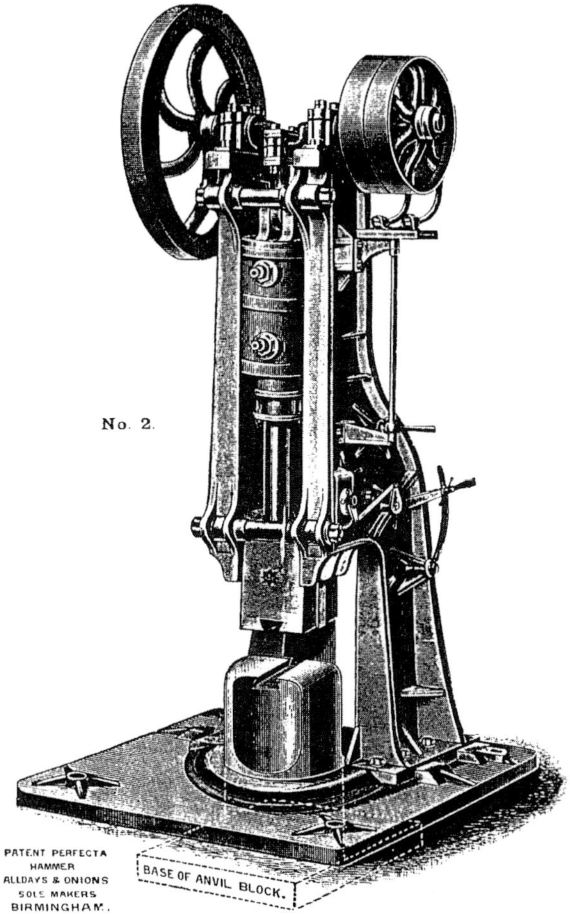

No. 2.

French double bick and double arch and sold by the hundredweight. Up to 1916 anvils were forged from a solid ingot by men who had got their working week down to 3 days and received a free supply of 8 pints of beer per day from the management, but when they demanded a further reduction in their working week to just 2 days Percy Allday decided enough was enough, sacked them all and thereafter produced anvils in cast steel.

Foundry equipment included geared wrought iron ladles in capacities from 10 cwts at £11-15s-0d each up to 12 ton types costing £94-17s-6d. Smaller, ungeared ladles, were made in a range from 5 cwts up to 2 tons and even smaller hand ladles were made from 28 lbs at 11s-0d each up to 56 lbs at 17s-6d each. Electric cupolas were available in sizes which produced 15 cwts to 10 tons of molten metal per hour, the largest cupola having a diameter of 5' 6", stood 28 feet high, weighed 9 tons 2 cwts minus its firebrick lining and cost £193. Just imagine what it would be like for a bricklayer to line the inside of one of these cupolas, which of necessity had to be done from inside, and using as little mortar as possible to prevent heat dissipation.

Stocks, dies and thread taps were made in Whitworth, Gas Tube and Brass or Copper Tube threads, together with hand powered screwing machines. A full set of Whitworth taps from 0.25" up to 2" complete with stocks in an oak case cost £47-7s-0d and screwing machines cost from £8 up to £28.

In 1900 a mechanical engineer with the company, Robert Chandler Jennings, took out a patent dealing with drop hammers of a type used in the forging industry which had also become part of the company's range of products. The Perfecta Patent range of pneumatic hammers was made in three basic types. The standard model in sizes from 2 cwts to 1 ton cost from £94 to £385, the largest model standing 14 feet 6 inches high. Overhang type hammers ranged from 28 lbs up to 15 cwts and were of more compact dimension than the standard models. Kick Stamp, Automatic Drop Stamp and Planishing hammers were also produced, being of much simpler construction than the pneumatic types.

Industrial fans of all sizes were also being produced and patented, including types for extracting hot gases from foundries, varnish works etc., and for mine and ship ventilation, some of these featuring water cooled bearings on the fan driving shafts. Ironically, these fans, despite all the other diverse products made by Alldays, were to become their only products in later years offering potential customers a wealth of experience in design, manufacture and installation in this highly specialised field of industrial extraction equipment. The mine ventilation fans were made in sizes from a number 84 which produced 23,600 cu.ft. of air per minute and required just over 11 h.p. to drive it, up to a number 340 model which produced 945,000 cu.ft. of air per minute and required no less than 1,050 h.p. to drive it. These huge fans required a clutch between the driving engine and the fan itself and Alldays were able to offer the "Western" hydraulic clutch made under Herschmann's patent which resembled internally expanding brake shoes in an open drum and operated by hydraulic cylinders which could be actuated remotely by an operator pulling a piece of string or wire.

Alldays Steel Plate Fan blowers were also made with, as the name implies, fabricated steel plate construction in capacities of 2,230 cu.ft./min to 14,240 cu.ft./min, requiring power units from 1 h.p. to 12.5 h.p. Ships fans were designed with their own

ILLUSTRATION

Two basic types of pneumatic power hammers were in production in 1900, the model on the left being made in 7 different sizes, 2 cwts, 3 cwts, 5 cwts, 7 cwts, 10 cwts, 15 cwts and 1 ton. The smallest model was 2ft 11ins wide between the inside of the vertical standards, stood 8ft high and cost £94. The 1 ton model was 5ft 1ins between the inside of the vertical standards, stood 14ft 6ins high and cost £385. These hammers were recommended for shovel plating, skelping scythes, steel tilting, shipyards and general smiths' work.

The other two models illustrated were of the overhang type, the No.2 model being offered in six sizes, 2 cwts, 3 cwts, 5 cwts, 7 cwts, 10 cwts and 15 cwts. The smallest model stood 8ft high and the largest model 12ft 1ins. These hammers were recommended for railway shops, dockyards, mills and general engineering shops.

The No.3 overhang type was available in five sizes, .25 cwt, .5 cwt, .75 cwt, 1 cwt and 1.25 cwts. These machines were 6ft high for the smallest model, with the rest standing 6ft 7ins high and they cost £55, £60, £70, £82 and £88 respectively. Obviously intended for much lighter work the No.3 model was recommended for file forging, table blade moulding, fork drawing, edge tool forging, cycle component forging and light smith work.

Interestingly, an Alldays drawing (Number 313) still exists for a massive 50 tons drop forging hammer designed in 1869, showing that, even then, the company was capable not only of designing such a massive piece of equipment, but was able to manufacture it!

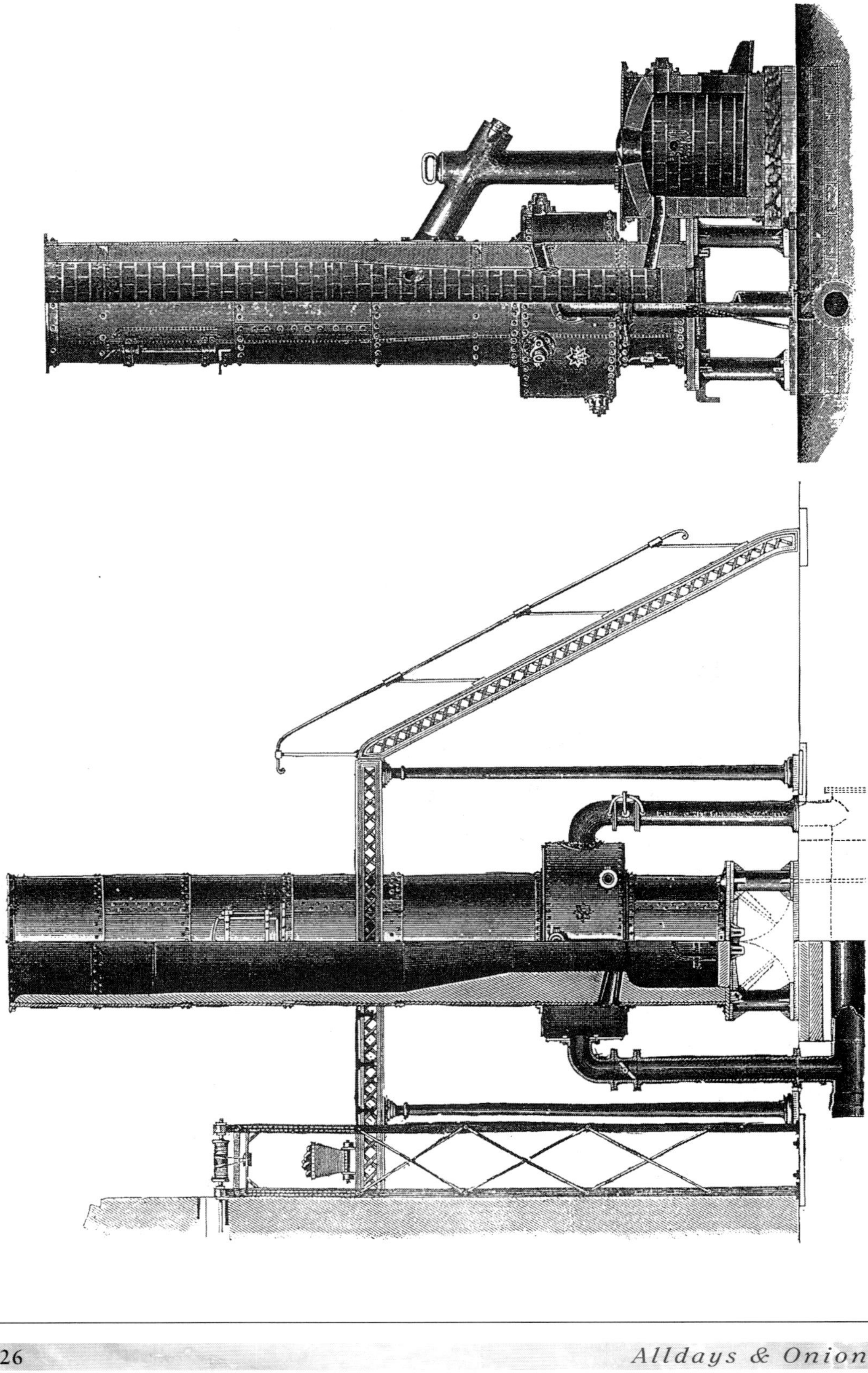

ILLUSTRATIONS

LEFT: As early as 1900 the company was building electric cupolas in a range of 9 standard models capable of producing 15 cwts to 10 tons of molten metal per hour. The smallest model was 2 feet in diameter with a 3 feet 3 inches diameter air belt, stood 15 feet high and weighed 1 ton 16 cwts, without its firebrick lining, and cost £43 plus £10 for the firebricks. The largest model was 5 feet 6 inches in diameter with a 7 feet 6 inches diameter air belt, stood 28 feet high, weighed 9 tons 2 cwts and cost £193 without its firebrick lining.

The cupolas were constructed of plate iron rivetted together with several rings of angle iron inside the shell to help support the firebrick lining. The staging/stairway to access the loading door and hoist to raise fuel and pig iron could also be supplied by Alldays.

The sectioned view of a cupola also shows a receiver attached to it which could accept metal as fast as it was melted and assisted the efficiency of operation of the cupola, and could also be used to store molten metal until required. Receivers were built in capacities of 10 cwts to 5 tons and cost £7 to £28 less the firebrick lining, according to size.

Although the company offered this standard range of cupolas and receivers it would also build larger models to customers specifications.

RIGHT: The tyre bending machines, pattern numbers A450 and A455, were used to roll steel strip, or angle, to form the steel tyres for vehicles, carriages or machinery. Pattern A450 could roll flat steel bars up to 4 inches by 0.875 inch and cost £6. A treble geared model able to roll steel bars up to 6 inches by 1 inch was also available at £9-10s-0d. Pattern A455 was similar to A450, but was fitted with special rolls for bending tee shaped, or angle iron and cost £7.

The rail bender and punch shown in the lower illustration was widely used by railway companies for bending and punching holes in rails up to 100lbs per yard in weight and could be supplied fitted to a trolley suitable for running on any gauge of railway track.

Tyre Bending Machine.

Pattern No. A 450.

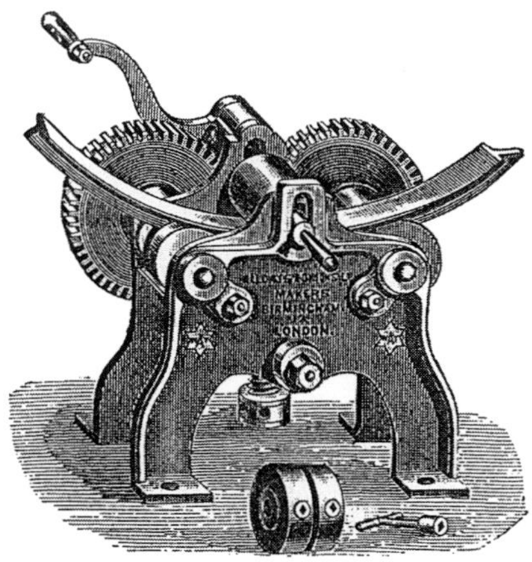

T&L IRON BENDING MACHINE.
Pattern No. A 455.

Alldays I.S.R. Pattern Rail Bender, or Punch combined.

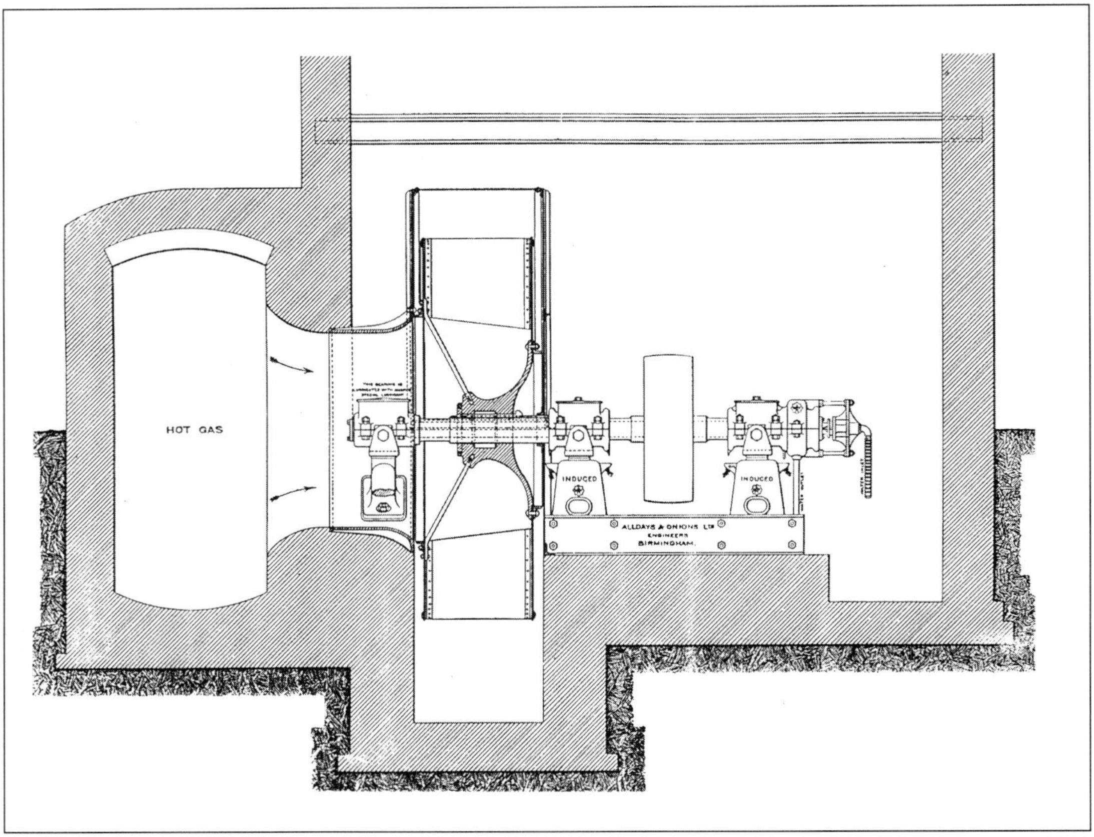
HOT GAS
INDUCED
INDUCED
ALLDAYS & ONIONS LTD
ENGINEERS
BIRMINGHAM.

ON HOUSE CHAMBERS
O, BUCKLERSBURY.
ALLDAYS & ONIONS
PNEUMATIC ENGINEERING Co LTD
WORKS-BIRMINGHAM.
ALLDAYS & ONIONS
PNEUMATIC ENGINEERING Co LTD
ESTABLISHED 1650
ALLDAYS & ONIONS
GREAT WESTERN
AND
MATCHLESS WORKS
BIRMINGHAM.
ALLDAYS & ONIONS
OA 721

integral steam driven 'engines' in a range from 3,660 cu.ft. per minute up to 20,740 cu.ft. per minute, the latter requiring over 28 h.p. to drive the fan. A special range of fans called the "Electro", directly driven by electric motors, was also in production in 1900, the motors being made 'in house' and costing from £28-10s-0d for a 450 cu.ft. per minute fan up to £75 for a fan delivering 3,900 cu.ft. of air per minute. The "Sydenham" range of pressure fans for forges, with cast iron casings, ran from a size A9 delivering 38 cu.ft. of air per minute up to a size B50 which delivered up to 16,050 cu.ft. of air per minute and required 78 h.p. to drive it. Root's patent blowers made under the Alldays "Climax" name were made for use with cupolas, blast furnaces, smiths fires and ventilating mines, the largest of these blowers, the size number 10 which could deliver 19,400 cu.ft. of air per minute was available with its own steam powered direct acting duplex engine which had a bore of 18½" diameter, stroke of 19", developed 75 h.p. and cost £550 complete with bedplate.

Alldays had obtained sole rights to manufacture Timmerman's patented simplex machine for bending, welding and stamping steel, a machine which was of value to any engineering firm.

Workshop machinery at this time was usually driven by overhead lineshafts, pulleys and leather belts and Alldays were able to offer all the necessary brackets, bearings, shafts and pulleys, the latter items being made in a range from 10" diameter and 4" wide at 14s-6d each up to 8' 4" diameter and 12" wide at £14 each. Fabricated from wrought iron these pulleys were said to be unbreakable, light, strong and with less friction absorbed less power to drive them, making them much more efficient than heavy cast iron type pulleys.

A large range of bench vices made in wrought iron and steel were available for such trades as general engineering, plumbers, silversmiths, blacksmiths and wire rope riggers as well as vices for general machine work. The smallest model was a number 4129 suitable for a home workshop at £1-2s-0d and the largest, a number 4136 riggers' vice more than 18" long cost £12-15s-0d. A range of portable and static vice benches was also made.

Bench drilling machines were mostly hand powered requiring the operator to turn a handle to power the drill bit and use his other hand to turn a horizontal wheel on top of the machine to provide the downward feed of the drill. These machines cost from £5-10s-0d up to £10-10s-0d, the latter being able to take a component up to 26" diameter and able to drill holes up to 1.75" diameter. Machines fitted with drive pulleys and with self acting feed could be had for about £1 extra.

Grindstones with trade names such as "Paragon" and "Ixion" were manufactured in sizes from 16" diameter up to 48" diameter in hand powered, foot treadle and power driven types. Heavier duty emery grinding machines were also in the company's catalogue with double grinding wheels up to 42" diameter by 4" wide and at a cost of £65. The grinding wheels themselves were advertised as coming from the company's own quarry, but its location is not known.

As well as manufacturing all this general and specialised engineering equipment the Great Western Works was also producing castings and forgings and carrying out all the machining of components for the ever increasing range of cycles, motor cars, motor cycles and commercial vehicles and must truly have been a hive of activity at this time.

ILLUSTRATION

This illustration from a 1905 catalogue shows a centrifugal fan in a typical installation, such as a foundry or steel works where hot gases needed to be extracted from the building. The large fan was mounted on a drive shaft supported by three huge bearings which were water cooled to prevent overheating. The drive pulley between the centre and right bearing could be driven from a gas or steam powered engine, or electric motor if electricity was available in the works.

An improved design of fan was patented in May 1906 and only when such designs are studied in detail can the complexity of these fans be appreciated and it was to be in this sphere of manufacturing that the Alldays name would be associated into the 1990s.

PHOTOGRAPH

Taken about 1908, this picture shows the London office and showroom of the company situated at Mansion House Chambers, 20 Bucklersbury. The picture itself would have made an excellent advertisement, as it shows a window full of bicycles, a tradesman's tricycle parked outside, together with a small motor car and light motor van parked in the gateway to Barge Yard at the side of the premises. From about 1907 the company did not attend the annual Stanley Cycle Show, preferring to exhibit their products in their own London showroom. (Photograph courtesy of Johnny Thomas)

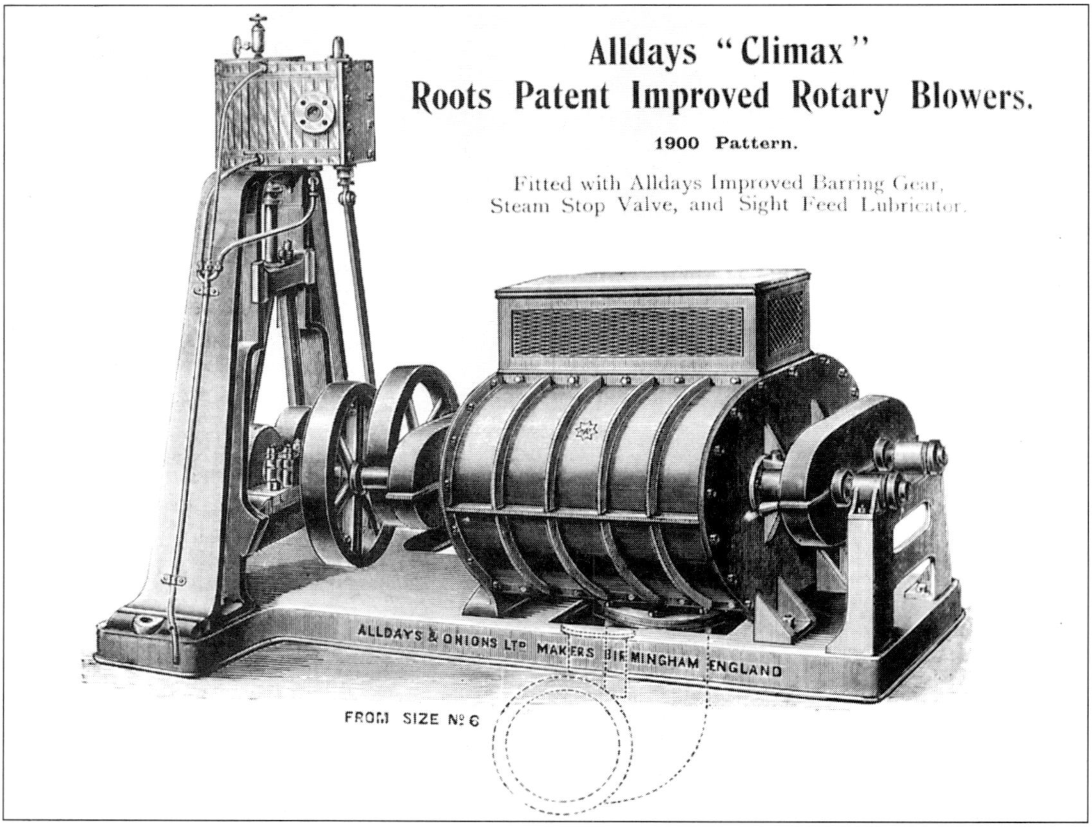
Alldays "Climax"
Roots Patent Improved Rotary Blowers.
1900 Pattern.
Fitted with Alldays Improved Barring Gear, Steam Stop Valve, and Sight Feed Lubricator.
ALLDAYS & ONIONS LTD MAKERS BIRMINGHAM ENGLAND
FROM SIZE No 6

DUPLEX
DUPLEX
ALLDAYS & ONIONS
MAKERS
BIRMINGHAM

PHOTOGRAPHS

TOP-LEFT: Roots Blowers were available in many sizes, especially to provide the blast of air required for blacksmiths' fires. In 1900 the largest of these blowers were available in a range to give 600 cu.ft./minute up to 19,400 cu.ft./minute and, like the example shown here, could be supplied on a bedplate complete with single cylinder steam engine to drive it. The engine shown here was known as an open-type where most of the working parts are exposed, but enclosed-type engines, like the one shown driving the fan in the illustration below, were also made as an option. The largest Roots blower built at this time was a number 10 model which could supply sufficient air for 220 fires and was driven by a 75 h.p. engine with a bore of 18.5" diameter, stroke of 19", ran at 245 r.p.m. and cost £550. The diameter of the delivery pipe, shown in dotted lines on this picture, would have been 26" in diameter.

BOTTOM-LEFT: "Duplex" exhausting fans, with cast iron casings, were made in sizes from 10.25" diameter up to 104.25" diameter, the example shown here being of the double inlet type of a kind supplied to varnish making works, etc. The largest fan in this range could extract over 20,000 cu.ft. of air per minute and cost £255 for the fan alone in 1900. The overall height of the largest fan was 10ft 2ins and width was 10ft 3ins. The overall length varied according to the type of drive employed, single pulley, double pulley, or directly as shown here, driven by an enclosed steam engine.

ILLUSTRATION

RIGHT: In 1884 William Allday junior took out patent number 15932 to cover this design of an attachment for a planing or shaping machine which enabled the 'valley' of a Roots type rotor lobe to be cut on the forward stroke of the machine and one half of each radiused end of the rotor lobe to be cut on the return stroke of the machine toolstock.

This attachment greatly facilitated the speed of machining and ensured much greater accuracy of the finished components than had been possible previously where several machine operations had been required to make the rotors used in these popular rotary blowers.

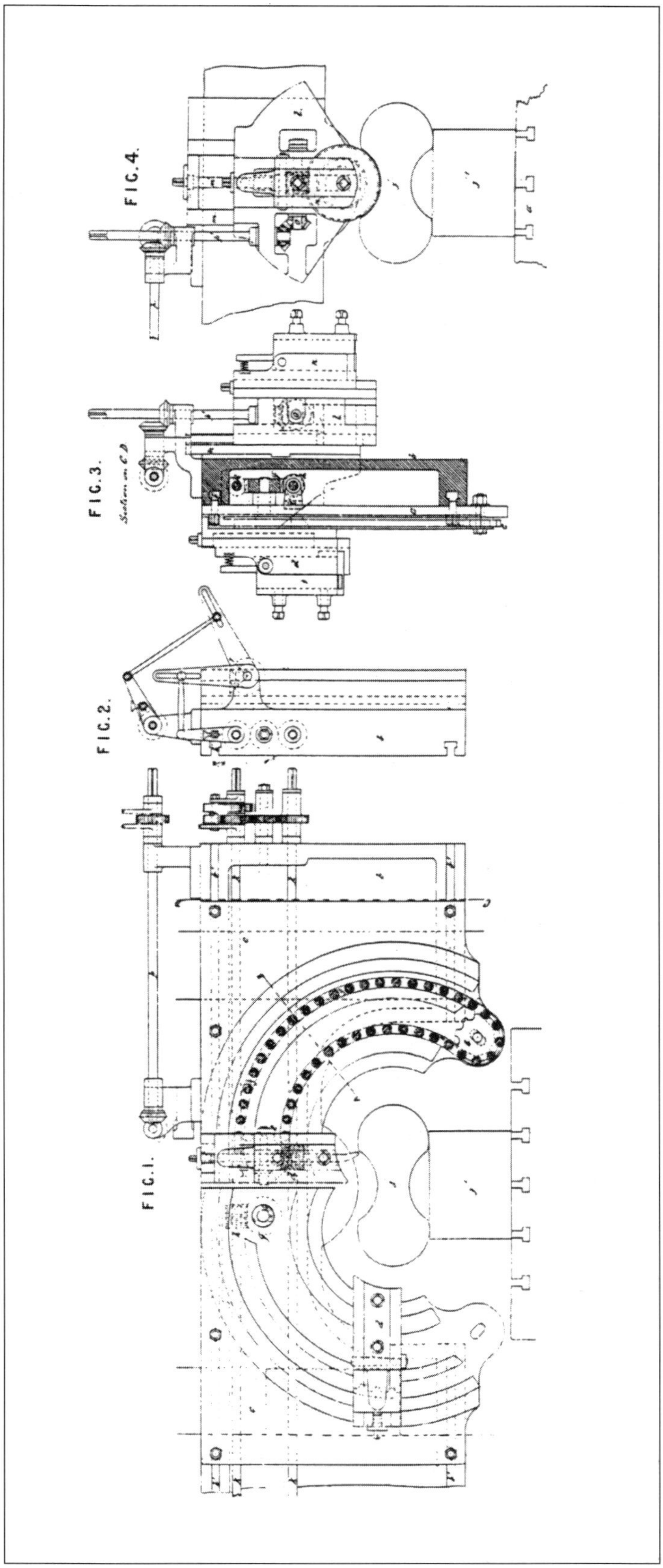

Boiler Makers' and Farriers' Tools.

Rivetters' Dollies. (Steel.)

1175

1177

1176

1178

Weights, 7 lbs. and upwards, per lb.

1179

Rivet Snaps. No. 1179.

Sizes to suit Rivets, Shanks $\frac{1}{4}$, $\frac{3}{8}$, $\frac{1}{2}$, $\frac{5}{8}$, $\frac{3}{4}$, $\frac{7}{8}$, 1 inch diameter.

Prices, per lb.

Farriers' Tools.

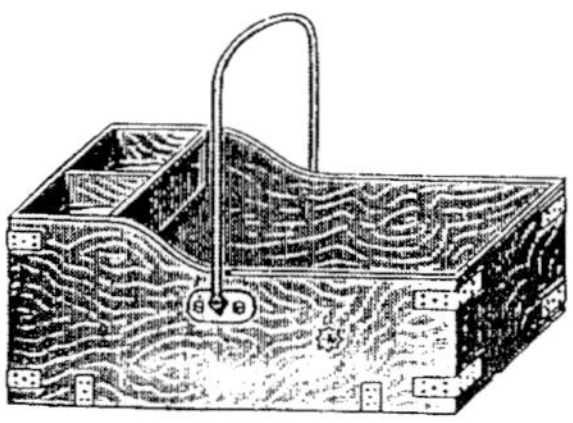
FARRIERS' TOOL BOX

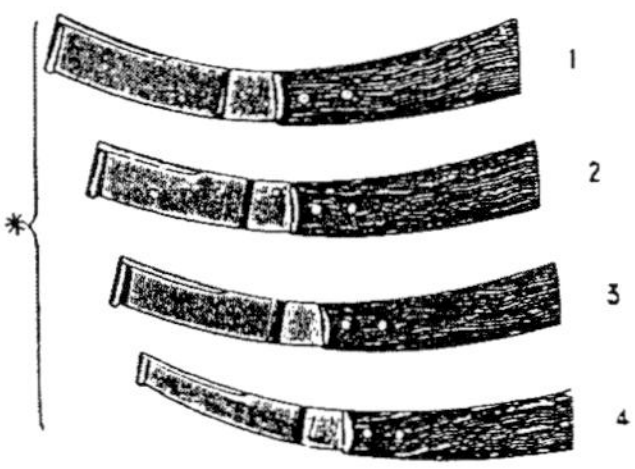

SET OF FARRIERS KNIVES

SHOE TONGS

SPRETCHEL

HORSE RASP

FLAT

HALF ROUND

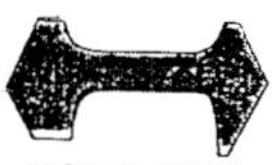
CLENCH KNIFE

STAKE

STAMP

SEAMER

SLEDGE HAMMER.

NAILING ON HAMR

STRAIGHTENING HAM*

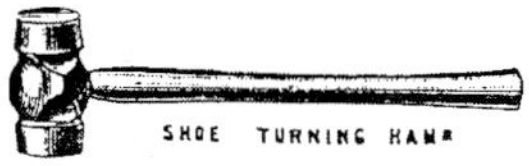
SHOE TURNING HAMR

Set of Farriers' Tools, as per above illustrations, complete with Box, net.

Marked * are best Sheffield make, but are included with those of our own manufacture to complete set.

Alldays & Onions Improved Triton and Eclipse Fog Horns

Captain Key's, R.N., Patent.

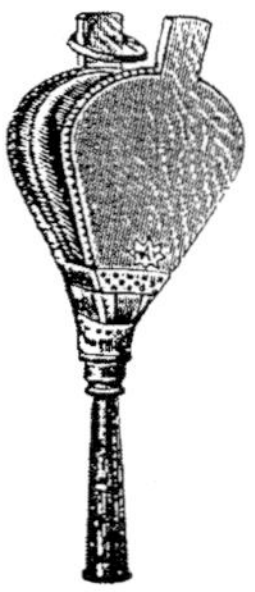
150

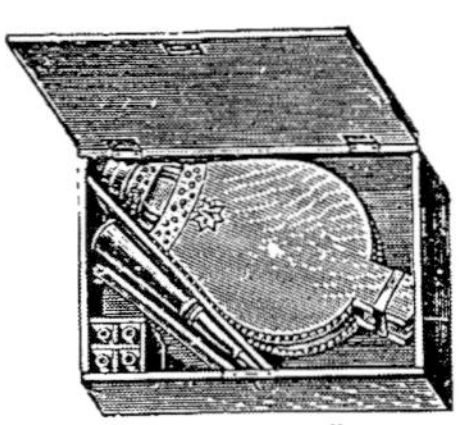
ALLDAY'S "TRITON" FOG HORN
BELLOWS Nº 1780

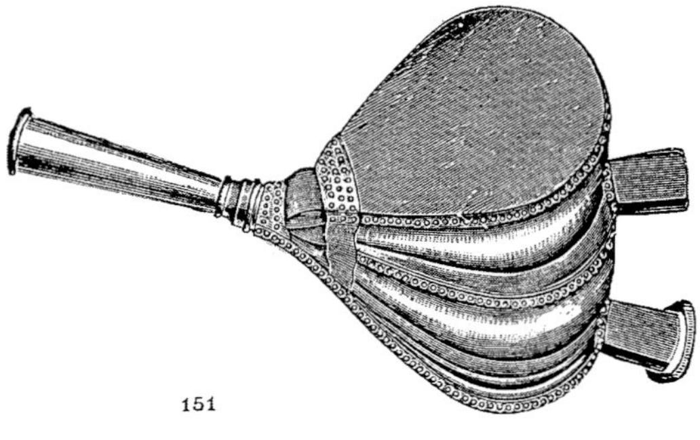
151

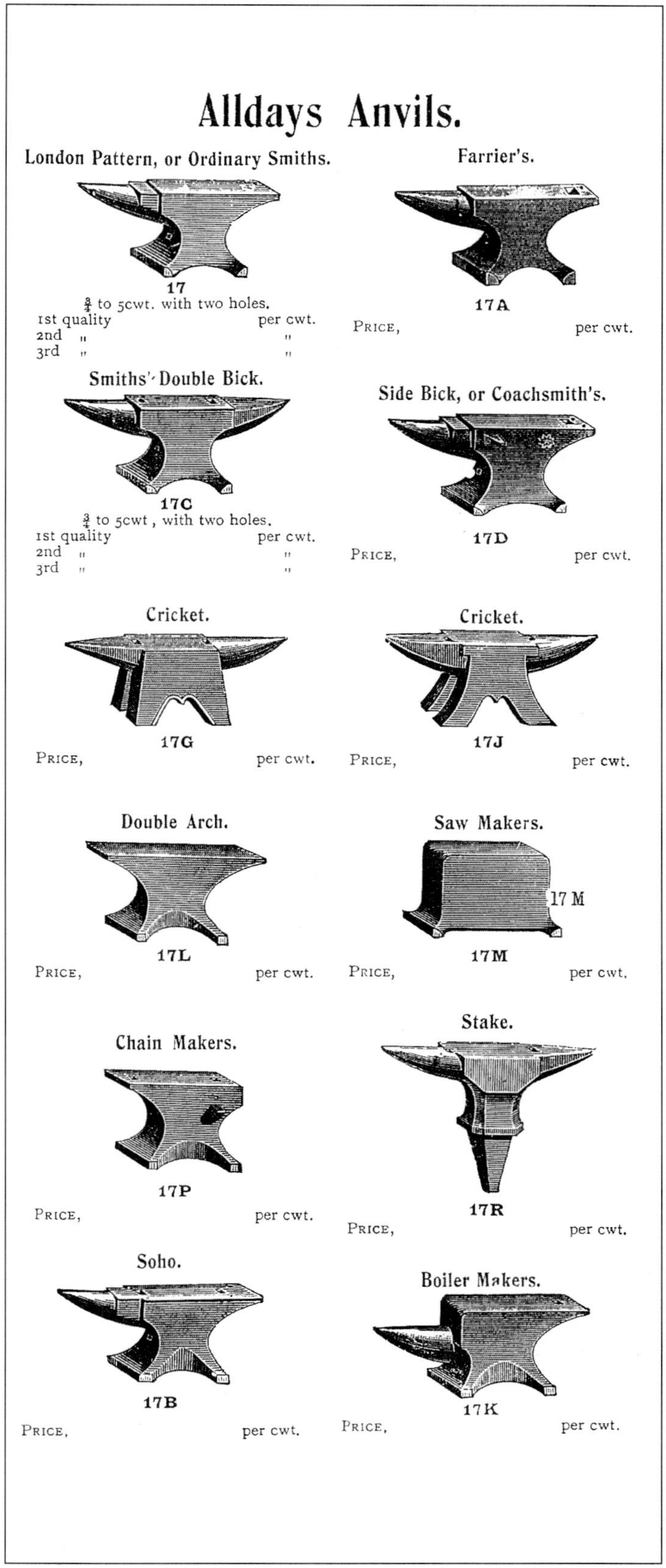

Alldays Anvils.

London Pattern, or Ordinary Smiths.

17

¾ to 5cwt. with two holes.

1st quality per cwt.
2nd " "
3rd " "

Farrier's.

17A

PRICE, per cwt.

Smiths' Double Bick.

17C

¾ to 5cwt , with two holes.

1st quality per cwt.
2nd " "
3rd " "

Side Bick, or Coachsmith's.

17D

PRICE, per cwt.

Cricket.

17G

PRICE, per cwt.

Cricket.

17J

PRICE, per cwt.

Double Arch.

17L

PRICE, per cwt.

Saw Makers.

17 M

17M

PRICE, per cwt.

Chain Makers.

17P

PRICE, per cwt.

Stake.

17R

PRICE, per cwt.

Soho.

17B

PRICE, per cwt.

Boiler Makers.

17K

PRICE, per cwt.

ILLUSTRATIONS

TOP LEFT: A small sample of some of the hand tools made by the company, in this case, for Boiler Makers and Farriers. The Rivetters Dollies numbers 1175 & 1178 cost 8d/lb and numbers 1176 & 1177 cost 10d/lb in 1905. The Rivet Snaps cost 1s-0d/lb. A complete set of Farriers Tools in a tool box as illustrated cost £1-12s-6d in 1884 and £2-11s-6d in 1905.

BOTTOM LEFT: These "Triton" and "Eclipse" Fog Horns in 1914 cost between £4-10s-6d and £5-19-6d for the "Triton" model and £9-0s-0d to £12-0s-0d for the "Eclipse" model. The fitted case cost 2s-6d or 3s-0d extra depending on the size. A "Norwegian Pattern" Fog Horn was also available, which looked like a square box with a handle to turn on one side, and cost £3 each.

ILLUSTRATION

RIGHT: Some of the different types of anvils made in 1900 which, at this time, were forged from a solid ingot of steel by men supplied with a free ration of beer each day because of their arduous working conditions.

In 1916 the manufacturing process for making anvils was changed from forging from a single ingot to casting in steel. The top face of the anvils were cast on to cold steel to provide a 'chilled' hard working surface, where previously the hardened top working face had been heat welded to the anvil.

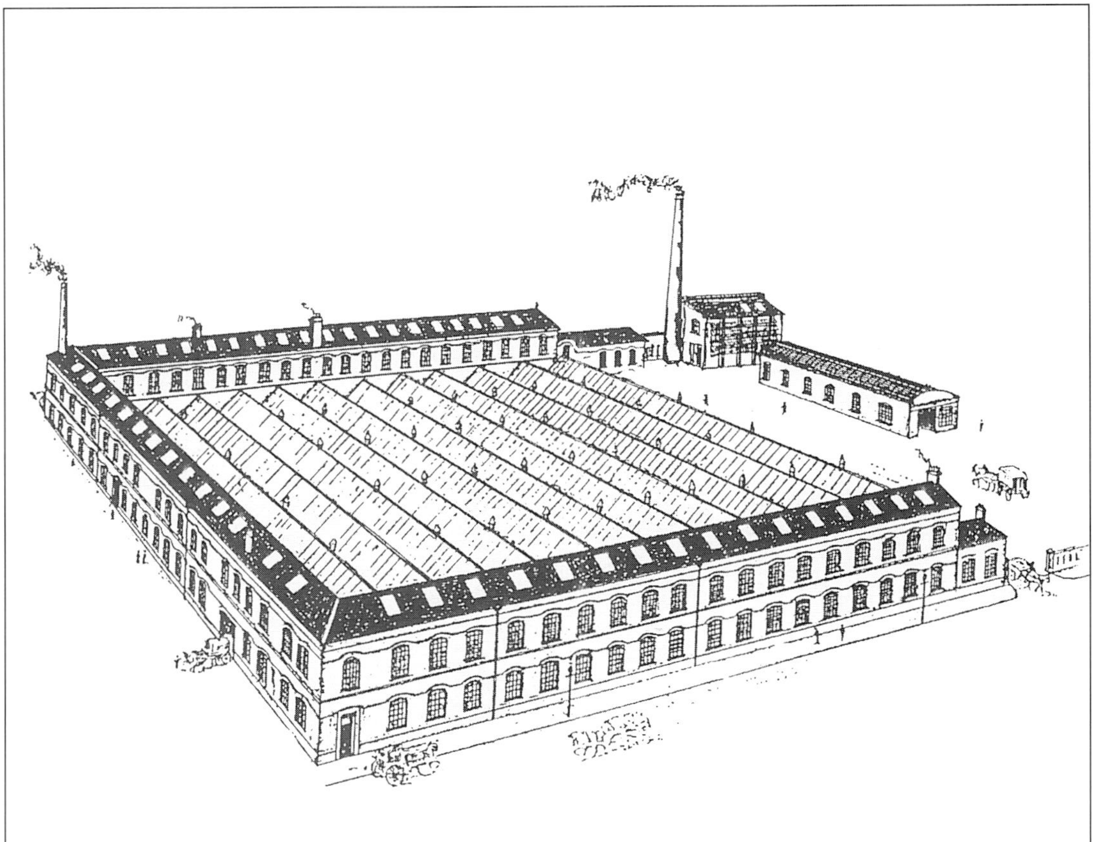

MATCHLESS WORKS

The company had begun to take a serious interest in the bicycle and although bicycles were already being built in many factories in Birmingham, including the B.S.A. factory in nearby Armoury Road, the Harry James factory in Sampson Road North and in the John Hanman factory in Sydenham Road itself, they decided to enter into cycle manufacturing themselves in a specially designed factory in Fallows Road. (Edmund Allday had taken out patents concerned with the design of velocipedes (cycles) as early as 1889.)

Plans were submitted in July 1896 for a two storey 'U' shaped building, the centre of the 'U' shape being a single storey workshop and the building having large open yards to two sides, the front of the building being built directly to the pavement in Fallows Road and the rear of the building being surrounded by recently constructed houses. (The drawings also included details for a second floor stores, which would have made the 'U' shaped building 3 storeys high, but the factory was actually built with only 2 storeys.) With the parent factory being situated so close to Fallows Road it appeared strange that the new factory, which quickly became known as the Matchless Works, even had its own Board Room. However, it was obviously the intention to make the new works as self contained as possible and wholly committed to the manufacture of cycles. (The choice of name was unfortunate, as "Matchless" cycles were already being made by the Midland Counties Cycle Company in Newport Pagnell.)

Production of cycles began as soon as the building was finished and suitably equipped and large orders were soon secured for cycles, tricycles and tradesman's cycles, the name of Alldays & Onions already being established as one synonymous with quality products. Advertisements for Alldays cycles appeared in 1896, even before the dedicated factory was built, and were worded in a typically Victorian manner. "For the Highest Possible Quality of Cycles that can be Produced, combined with the Very Best Value Money can buy" and "For all round Excellence, combined with a Moderate Charge, You cannot select a Better Cycle than — Alldays", were two typical examples. The cycle was really in its heyday at this time with specially designed clothing being produced and carriers for 'cycle-cameras' and one establishment was even advertising a handgun, suitable for a cyclist to carry in case of attack by a highwayman!!

In December 1896 the company reported that cycles made from their patented corrugated steel tubing had proved so popular that they had been unable to supply the demand. At the Stanley Cycle Show that year one cycle from the huge order received from the British South Africa Company was shown resplendent in its red enamelled finish and complete with attachments to carry a rifle and sword!

In October 1897 some temporary cart sheds were erected in the open yard to the south of the works and in December 1905 the open yard to the north of the works, which measured 66 feet by 156 feet, was partly roofed over by an open ended two bay building 66 feet by 128 feet. The building comprised one bay 44 feet wide and one bay 22 feet wide. Such was the slope of the ground from the pavement in Fallows Road to the back of the works that the building was 10 feet 9 inches high to the eaves close to the road and 14 feet 9 inches high to the eaves at the back of the site, the overall height of the building being 20 feet.

ILLUSTRATION

This artist's impression of the Fallows Road Matchless Works appeared in company sales literature about 1908. Whilst giving some idea of how the works appeared in 1896/7 it contains several errors. The land on which the factory was built descended sharply in both directions from the front corner of the building in the left foreground, but this is not shown in the drawing. The original picture also shows a pavement running along the left hand side of the building, as if the factory stood at the corner of two roads. In fact, it was 200 yards from the corner of Anderton Road and the area to the left of the building was an open yard.

The roof detail above the single storey area making up the centre of the 'U' shaped building is totally incorrect, as it comprised five ridges from the right hand side of the building, then a single wide span divided into two sections above the area of the works 'H', 'L', shown on the block plan elsewhere in this chapter.

The chimney at the far left hand corner was from the heating chamber 'A' located in the basement of the building and the small single storey building on the far right at the front of the works was the engine house 'R'. The doorway at the left hand front corner of the building accessed the works from the roadway and the first floor offices via a stairway 'Q', 'q'.

PHOTOGRAPH

Although described in company literature as a machine shop, this view, taken in the Matchless Works about 1908, clearly shows an area of the works devoted to inspection and assembly. On the first row of work benches aluminium castings are being assembled and on the second row of benches bicycles are being assembled with the bottom of the frames held in bench vices. Of the machines visible in the picture most are lathes to the right hand rear side of the building and to the left can be seen a power press and some bench mounted drilling machines. (Photograph courtesy of Johnny Thomas)

"MATCHLESS WORKS"

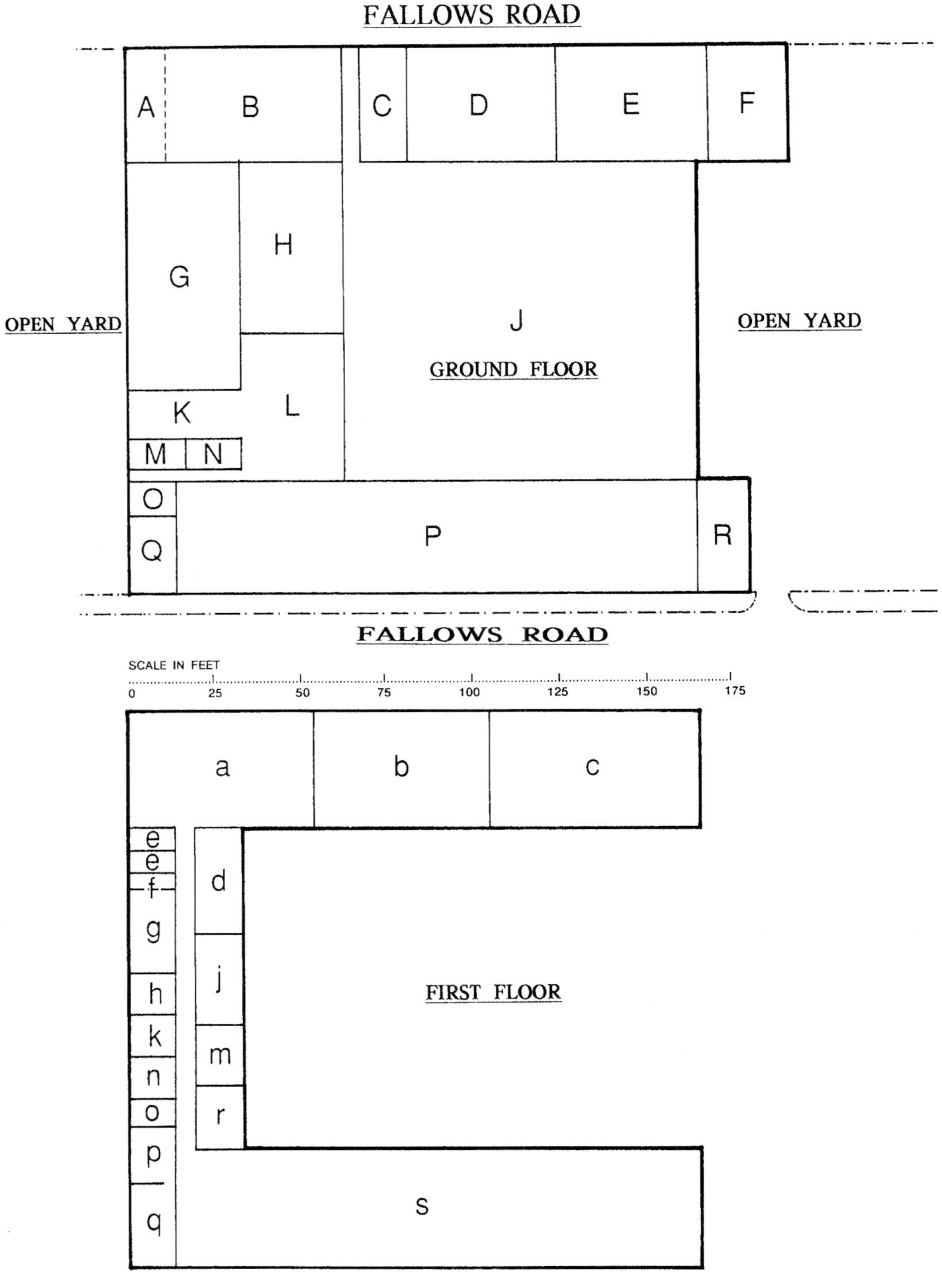

In February 1897 the company announced its intention of floating the cycle works on the Stock Market for £75,000, but in November they decided to cover the outlay for the new Fallows Road works by issuing Debenture Bonds to the value of £80,000. At the Stanley Show that year, held at the Royal Agricultural Hall in Islington, a very elegent ladies' cycle was shown with a silver lined frame and pearl handlebars with solid silver mounts. At the following year's show 25 machines were exhibited, including their "Imperial" Road Racer, and the company had reported profits of £18,301-8s-7d for 1896/7 and £10,174-6s-7d for the year 1897/8 with 6% dividends being paid on Preference Shares and 10% on Ordinary Shares.

The War Office gave Alldays their first large government contract in 1898 which was rapidly followed by four more contracts and in February 1901 by the largest contract to be awarded to the company so far, for 800 cycles. More than 40 of the company's cycles were used by despatch riders in the South African Boer War and a rickshaw based on their cycle components was supplied to His Majesty the Sultan of Morocco. The General Post Office were to buy Alldays cycles in large numbers and the various types of cycles were soon given names such as "Alldays Matchless", "Lady's Enchantress" and "Imperial Roadster".

At the 1899 National Cycle Show Alldays had a display of 21 safety cycles, a tandem, a carrier tricycle and a tradesman's handcart and on a separate stand, a motorised quadricycle powered by a rear mounted 2.25 h.p. De-Dion engine. The company announced a net profit of £11,245-9s-6d for the year 1898/99 with dividends of 6% on Preference Shares and 10% on Ordinary Shares and the shareholders agreed to vote the sum of £500 to the University of Birmingham. At the shareholders' meeting held in December 1900 a profit of £10,233 was announced with dividends of 5% on Preference Shares and 10% on Ordinary Shares. The cycle works, it was reported, had made more cycles that year than any time previously, but less profit had been made, due to the ever increasing competition which had sent some manufacturers into liquidation. However, the 'motor' side of the business was reported to be doing well. The two seater Alldays "Traveller" Voiturette had been introduced fitted with a 2.75 h.p. De-Dion engine and a two speed Champion clutch. This machine was priced at £120 and a larger 3.5 h.p. engine was planned for later models. At the Stanley Show that year, as well as the new motor car, cycles were shown with free-wheels and brakes (cycle braking until this time had relied upon 'back-pedalling', but various kinds of brakes were now coming on the market and Alldays patented their own design, which acted against the front tyre, in patent number 21950 dated November 1893) on such models as the "Imperial Roadster", "Imperial Lady's Enchantress" and the oddly named "Corrugated Road Racer". The latter machine was still using the corrugated steel tubing patented by the company in 1893, but this was rapidly being overtaken by the use of conventional steel tubing which had been adopted by all the major cycle manufacturers. (Patent number 21950 covered "Improvements in Bicycles, Tricycles and other Velocipedes" which included brakes, chain sprockets, wheel spindles, joints to connect frame tubing and various aspects of the frame design associated with a tandem.)

The trend towards lighter cycles also had its influence on Alldays machines in 1901 when they introduced their "Featherweight Imperial" which weighed 26½ lbs and cost £10-10s-0d and a lighter

DIAGRAM

BLOCK PLAN OF THE "MATCHLESS WORKS" FALLOWS ROAD - 1896

GROUND FLOOR

A. Heating Chamber 30ft x 12ft located in the basement.
B. Stores 30ft x 60ft.
C. Engine House 30ft x 14ft.
D. Polishing Shop 30ft x 42ft.
E. Brazing & Case Hardening Shop 30ft x 42ft.
F. Smiths Shop 30ft x 22ft. (Single storey building.)
G. Packing Room 30ft x 66ft.
H. Stores 30ft x 48ft. (Single storey building.)
J. Machine Shop 90ft x 100ft. (Single storey building.)
K. Cartway 12ft wide leading from open yard to covered yard 'L'.
L. Covered Yard 30ft x 40ft. (Single storey building.)
M. Office 9ft x 14ft.
N. Time Office 9ft x 14ft.
O. Experimental & Pattern Shop 10ft x 12ft.
P. Machine Shop 30ft x 150ft.
Q. Entrance from Fallows Road and stairway to first floor.
R. Engine House 30ft x 14ft. (Single storey building.)

FIRST FLOOR

a. Plating Shop 30ft x 52ft.
b. Wheel Shop 30ft x 50ft.
c. Enamelling Shop 30ft x 60ft.
d. Show Room 13ft x 30ft.
e. Toilets.
f. Safe 4ft 6ins deep built into the wall of the office 'g'.
g. General Clerks' Office 12ft x 22ft.
h. Private Office 12ft x 10ft.
j. Board Room 13ft x 25ft.
k. Private Office 12ft x 10ft.
m. Travellers' Room 13ft x 16ft.
n. Cashiers' Office 12ft x 11ft.
o. Enquiry Office 12ft x 8ft.
p. Waiting Room 12ft x 15ft.
q. Stairway from ground floor.
r. Typewriting Office 13ft x 18ft.
s. Finishing Shop 30ft x 150ft.

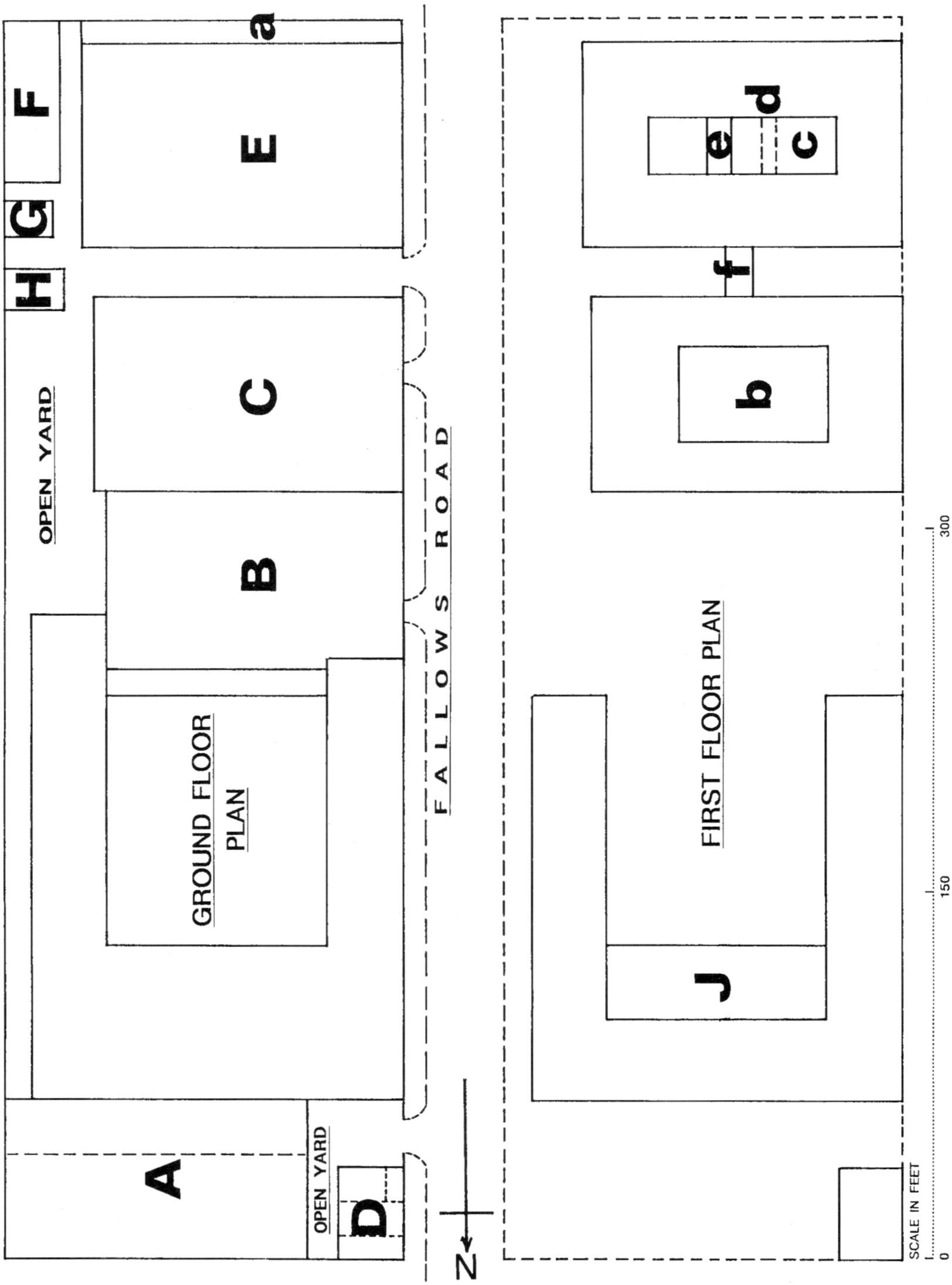
OPEN YARD
F
G
H
E
a
C
B
GROUND FLOOR PLAN
A
OPEN YARD
D
FALLOWS ROAD
N
d
e
c
f
b
FIRST FLOOR PLAN
J
SCALE IN FEET
0
150
300

BLOCK PLAN

The block plan opposite shows how the Matchless Works in Fallows Road was developed between 1905 and 1910. During this period some temporary buildings had been erected, repositioned and finally removed from the site. No planning applications have been found for some of the smaller buildings.

A. Most of the open yard to the left of the original works was roofed over in 1905 by this building which was completely open at the front to give access from the remaining part of the open yard. The building was built in two bays, one 44 feet wide and the other 22 feet wide. The building was 128 feet long.

B. This single storey building was erected in 1906 in four bays and a passageway was left between it and the original building where there was a 3 feet difference in the ground levels. The overall dimensions were 70 feet wide at the roadway end, 80 feet wide at the opposite end and 120 feet long. 9,240 square feet were added to the production space by the erection of this building.

C. In December 1906 plans were drawn up to build this two storey warehouse 82 feet wide by 130 feet long which had a central well 'b' to provide natural lighting to the first floor storage areas.

D. A Caretaker's House was built in the open yard in 1907, the ground floor of which comprised 3 bays for storing motor cars and a Lobby with a staircase leading to the first floor which provided the living quarters for the factory caretaker.

E. This 3 storey building was erected in 1910 with a 21 feet wide cartway between it and building 'C'. The building was 84 feet wide at the roadway plus a 10 feet wide covered passageway 'a' to the right and it was 134 feet long, adding 30,700 square feet of production space. On the first floor plan the open well 'c' in the centre of the building can be seen, together with the lift 'e' and a 4 feet wide gangway 'd' across the open well. A gangway 'f' 12 feet 9 ins wide connected the first floor of this building to the first floor of building 'C'.

F. Wood Store with open front, the building measured 25 feet by 67 feet.

G. Shed 15 feet by 20 feet.

H. Shed 16 feet by 25 feet. (No planning applications were found for any of these sheds.)

J. This Store Room was built in 1910 above the ground floor store and covered yard of the original building and was built with each of the long sides sloping inwards to allow light into the existing offices on the first floor.

version of the "Matchless" machine was also put into production. At the National Cycle Show that year 29 cycles were exhibited and one of the exhibits at the 1902 show was a military cycle used by a Lieutenant Carus Wilson during the Boer War. The "Traveller" motor car had sold well at the show and one jounalist described it as "a capital little vehicle".

At the shareholders meeting in November 1901 the company announced a profit of £10,657-17s-11d and a further £500 was voted to the Birmingham University Fund. The following year, 1902, profits fell to £7,850-14s-4d, but Mr William Allday disclosed to the meeting that during the previous 12½ years the company had written off £44,208-18s-7d in depreciation of plant and machinery and, as well as paying good dividends to shareholders, had managed to accumulate £20,000 in their reserve fund. Much of the reduction in the current year's profit had been caused by falling demands in general engineering work, but the cycle and motor side of the business had actually gone into profit. The Post Office ordered 500 cycles in 1902, some with trailers to carry mail, and in November 1903 placed an order for 600 "Matchless" cycles and this was followed by an order for 800 cycles in April 1904.

The company's "Imperial" cycles for men and women cost £12-12s-0d and now featured Eadie two speed gears. At the National Cycle Show in 1903 Alldays had their largest display which featured 30 cycles and motor cycles. However, profits had fallen again to £6,657 illustrating just how competitive the cycle and general engineering industry had become.

By 1903 the Alldays "Traveller" voiturette was in production powered by a 4 h.p. single cylinder engine which featured an air cooled cylinder and water cooled cylinder head. This little car was available with either 2 or 3 seats and had a two speed gearbox and wheel steering. Whilst the castings for the petrol engines and gearboxes were produced at the Great Western Works all the assembly of motor cars was to be undertaken at the Matchless Works.

In 1904 the price of the "Imperial Roadster" had been reduced to £10-10s-0d and the "Matchless Roadster" now cost £8-8s-0d, both machines featuring Clincher 'A1' tyres, Crabbe pattern brakes and with Alldays own two speed gears available as an option for £2-2s-0d extra. The British cycle industry had exported machines to the value of £584,000 in 1902 and £707,000 in 1903 with £133,000 worth of cycles being imported in 1902 and £84,000 worth imported in 1903.

In February 1904 John Collingwood Onions died at the age of 62 years. John Onions had been a member of the Yardley School Board and had represented the district of Yardley in the Worcestershire County Council.

For the 1905 season the company introduced up-market versions of the "Imperial Roadster" and "Imperial Lady's Enchantress" cycles fitted with Eadie two speed and back pedalling hubs and best quality tyres for £13 and entered the budget market with men's cycles called "King Edward" and lady's cycle called "Queen Alexandra" at £6-6s-0d and £6-10s-0d respectively. In March 1905 the GPO ordered 100 carrier tricycles followed later in the year by an order for 1,200 "Matchless" cycles and the War Office ordered a further 300 military cycles.

In May 1906 the Fallows Road factory was extended by the erection of a single storey building measuring 70 feet by 120 feet and built in four long bays which were about 3 feet below the existing two

You can go over Every Part

WITH A **MICROSCOPE**

AND FIND EACH DETAIL OF THE

ALLDAYS

Cycles

ABSOLUTELY PERFECT.

There's No Shoddy Work

Covered by Thick Plate and Enamel.

EXAMINE THEM CAREFULLY AT OUR STAND!

224, NATIONAL SHOW,

ALLDAYS & ONIONS Pneumatic Engineering Co., Ltd.,

Matchless Works, BIRMINGHAM, and

15, Sherbourne Lane, LONDON, E.C.

storey building. Described as a Fitting Shop the building was 16 feet high to the eaves and 25 feet high to the ridges of the asymmetric roofs. This particular building is thought to have been used later, certainly by 1909, as a Body Makers' Shop. The architect for this project was Messrs Riley & Smith of Kings Court, Colmore Row.

In December of 1906 plans were drawn up by Ewen Harper & Bros., to add a large warehouse to the works and to build a house for the factory caretaker. The 2 storey warehouse was 82 feet wide by 130 feet long and built partly on the existing walls of the recently erected Fitting Shop. An open well in the centre of the first floor provided natural lighting to this floor of the building which added 18,920 square feet of storage area to the works.

The Caretaker's House built in the corner of the open yard to the north of the site had only a Lobby on the ground floor with access to the roadway and the yard, together with a staircase leading to the house proper on the first floor. (Beneath the house were 3 bays for storing motor cars.) This building measured 27 feet by 37 feet with a ground floor height of 11 feet and first floor height of 9 feet 6 inches, the overall height of the building being 31 feet. The house was provided with 3 bedrooms, a living room, parlour, kitchen, bathroom and pantry.

The GPO ordered 700 Alldays cycles in 1906 and a further 1,200 in 1907, these orders being secured at a time of increasing competition, but showing that these machines had achieved a reputation for good value and reliability. (The GPO operated 400 cycles and 100 carrier tricycles in the Greater London area alone, the cycles chiefly being used for carrying messages by boys recruited between the ages of 13-15 years old and paid 7 shillings per week. The carrier tricycles which weighed about 130 pounds were employed mainly for collection of mail and distribution of mail bags between sorting offices, the cycles being able to carry a load of about 1 cwt.)

The company did not attend the 1907 Stanley Show, preferring to invite customers and dealers to their showroom at number 20 Bucklersbury, London. Here practically the whole range of cycles was on show and included Tradesmen's cycles with front and rear carriers at £7-10s-0d, a top of the range Tradesmen's cycle, the "Express Deliverer" at £18-0-0 and ordinary "Roadster" cycles from £6-10s-0d to £13-0-0. Juvenile safety cycles had been introduced priced at £6-6s-0d and "Road Racers" were available at £6-6s-0d and £8-8s-0d and tricycles were priced from £18-0s-0d up to £21-15s-0d.

During 1908 the company had doubled the output of cycles from the Matchless Works where the workforce had worked an average of 53½ hours per week throughout the year to keep up with the demand for their products. Profits for the year 1908-1909 had reached £22,822 and after paying dividends of 6% on preference shares and 5% on ordinary shares a total of £17,947-15s-1d was carried forward.

On July 7th 1909 King Edward and Queen Alexandra visited Birmingham to officially open the new university buildings in Edgbaston and the city centre streets were decorated for the occasion. In New Street, close to Needless Alley, a semi-circular arch 26 feet wide and 22 feet high was erected across the road by the Cycle Trade which was covered with 8,000 cycle parts and carried two greetings for the royal visitors, "Cycle Trade Welcomes Their Majesties" on one side and "Long Life To Their Majesties" on the reverse side. Costing £300 the arch had been paid for by donations

ILLUSTRATION

One of the earliest illustrated adverts for Alldays cycles was this example which appeared in the November 20th 1898 edition of "Cycling" magazine. The company had been making cycles for just two years when this advert appeared, but they were obviously very proud of the workmanship which went into all their products.

My 'Alldays Enchantress' is such a lovely little Bike
ALLDAY'S CYCLES
ARE THE
WORLD'S BEST
CYCLES.
THE
WORLD'S
BEST
CYCLES.
SEND FOR
CATALOGUE.
AGENTS WANTED WHERE NOT
. . . . REPRESENTED
ESTABLISHED 1650.
ALLDAYS & ONIONS, PNEUMATIC ENGINEERING CO., LTD., BIRMINGHAM.

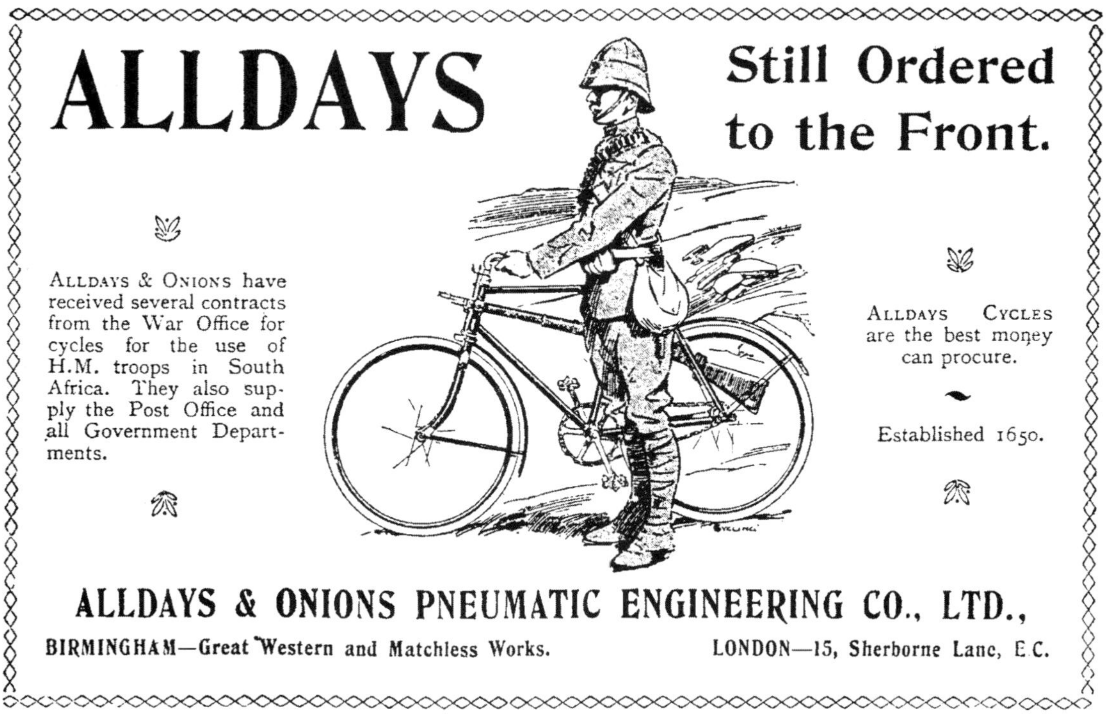
ALLDAYS
Still Ordered
to the Front.
ALLDAYS & ONIONS have received several contracts from the War Office for cycles for the use of H.M. troops in South Africa. They also supply the Post Office and all Government Departments.
ALLDAYS CYCLES are the best money can procure.
Established 1650.
ALLDAYS & ONIONS PNEUMATIC ENGINEERING CO., LTD.,
BIRMINGHAM—Great Western and Matchless Works.
LONDON—15, Sherborne Lane, E.C.

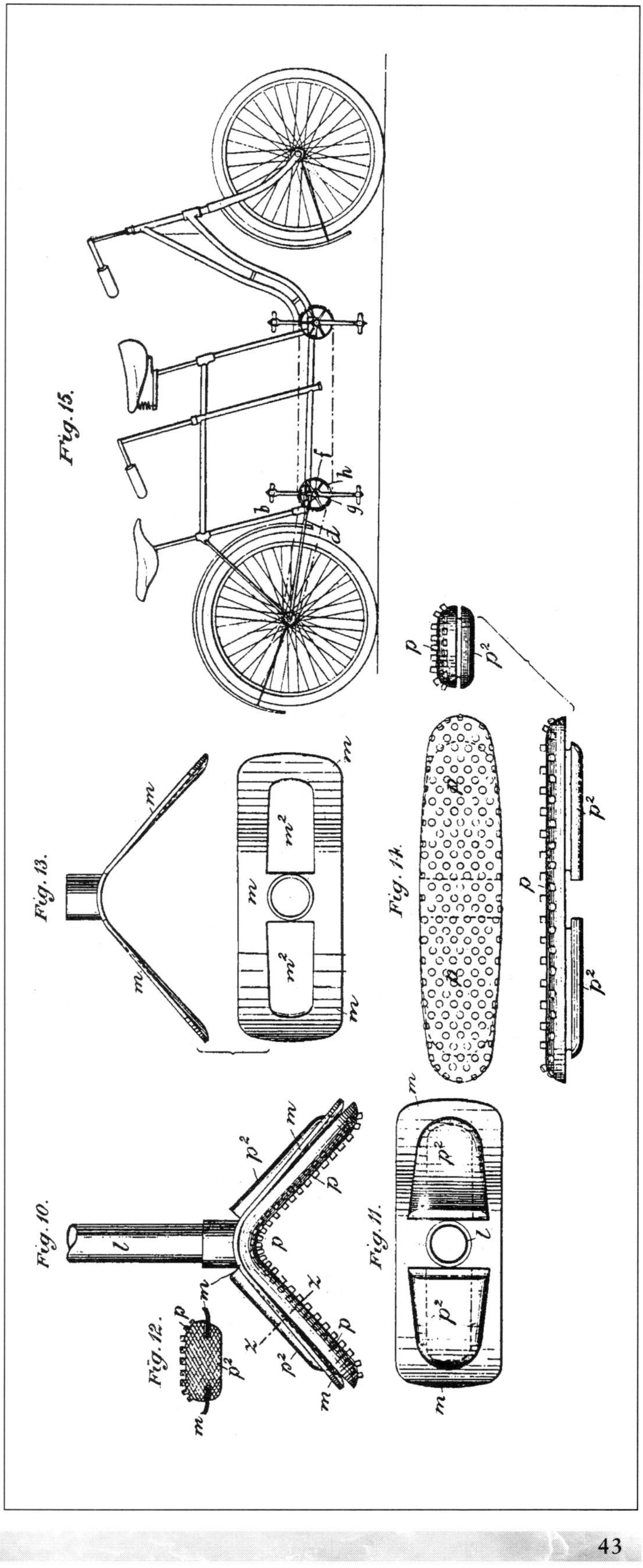

ILLUSTRATIONS

LEFT: These illustrations show two 1899 advertisements from "Cycling" magazine. The upper advertisement is an excellent example of 'art in advertising' which was at its zenith between the end of the 19th century and the 1920s.

Advertising the "Alldays Enchantress" lady's cycle, the advertisement leaves no doubt in your mind who made the World's Best Cycles!

The lower advertisement illustrates the use to which Alldays cycles were put during wartime activities in South Africa where many cycles were used not only to deliver messages, but to provide welcome transport to infantrymen.

RIGHT: Various diagrams from patent number 21950 taken out by William and Edmund Allday in 1893 for improvements to "Bicycles, Tricycles and other Velocipedes". Figures 10 to 14 show the component parts for a brake assembly which acted against the tyre of the bicycle and Figure 15 shows Alldays idea of how a tandem should have looked, with the lady at the front and the gentleman seated to the rear.

Tricycle.
"ALLDAYS CYCLES"
BIRMINGHAM
A&O.

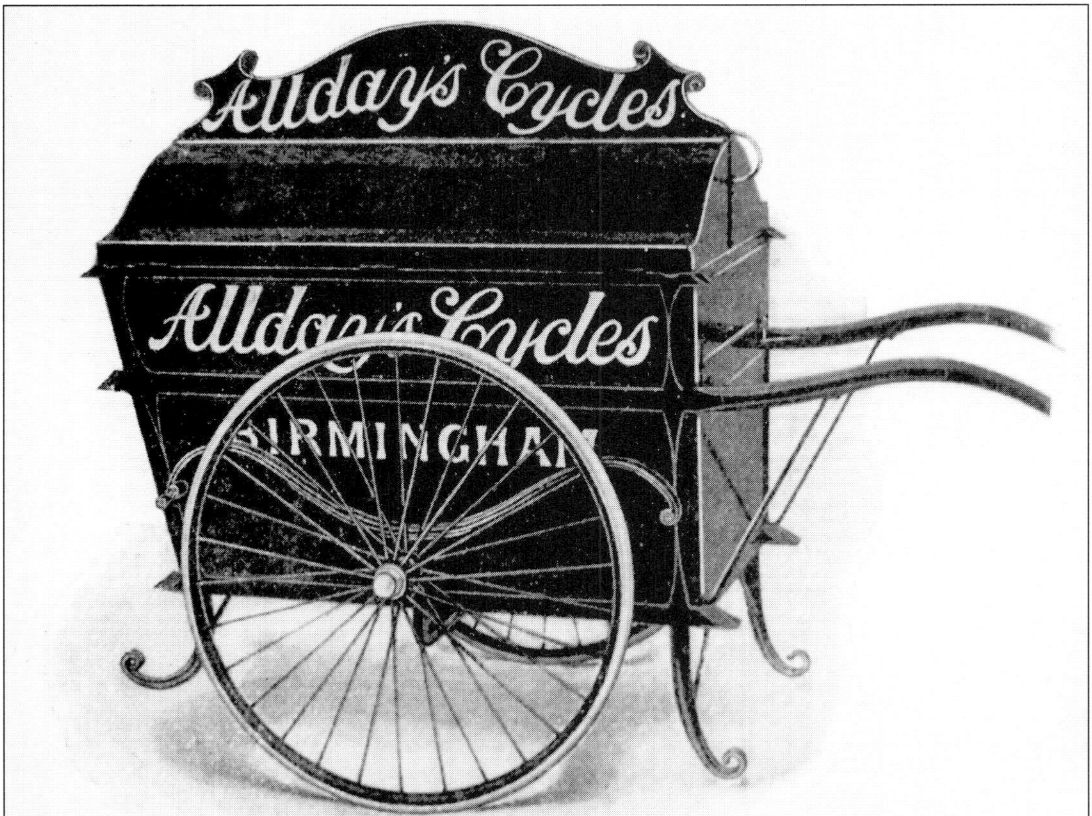
Allday's Cycles
Allday's Cycles
BIRMINGHAM

from companies in the cycle and associated industries and Alldays had contributed £10 towards the cost.

At the National Cycle Show that year an improved version of the "Imperial Matchless" Roadster was introduced priced at £10-15s-0d and a new box carrier tricycle was shown with their recently patented design of frame. Racing cycles on display included the "Matchless", "Imperial" and "Popular". Among the exhibits at the following year's show was a GPO box carrier tricycle fitted with the new Clincher "Salvus" tyres and a novel solid tyred rickshaw complete with Cape Cart Hood, built for an undisclosed customer.

After some over-production and decline in the sales of cycles it was heartening to the British industry to discover that in the first 6 months of 1910 exports had brought in £1,055,323.

In August 1910 alterations were made to the Matchless Works when the roof was removed from above the ground floor Store(H) and Covered Yard(L) shown in the Block Plan, and a two storey Store Room was built 26 feet high to match the roof line of the existing building, but the store room itself had sloping glazed sides which allowed light into this extension and the offices (d, j, m and r) inside the existing upper floor of the original building. Planning permission was sought in October 1910 for a four storey building for the Matchless Works to be built at the southern end of the site immediately behind houses along Walford Road. The building was 84 feet wide, plus a 10 feet wide covered passageway on the ground floor, and 168 feet long. The upper floors had a glazed centre well 20 feet wide by 55 feet to allow light into the inner parts of the building and this well also contained a lift 10 feet by 20 feet which served all floors and a four feet wide gangway ran through the well at each floor level. A gangway led from the first floor of this building, across the 21 feet

PHOTOGRAPHS

LEFT: Two illustrations taken from a 1900 general catalogue. The upper picture shows a Tradesman's Carrier Tricycle which would have been used by all sorts of busy retailers for delivering goods, and also by the GPO for moving mail bags between sorting offices.

The lower picture shows a Tradesman's Handcart which used some cycle parts in its construction, but with solid tyres was obviously used to carry heavier loads than the tricycle and would have been used not only to transport goods from place to place, but more probably as a cart from which goods could be sold in streets and markets.

RIGHT: A view of the Chassis Erecting Shop in the Matchless Works, about 1909. Up to 40 chassis could be built here at a time, using the simple wooden trestles on which chassis were mounted whilst assembly work continued to the point where chassis were complete with all the parts to make a running chassis suitable for road testing.

ALLDAY'S TRAVELLER VOITURETTE (Registered).

4 h.p. Motor water-cooled combustion chamber). | Speeds 4 to 20 miles per hour, | No belts, no chains. | Simplest, the best. | Send for catalogue.

ALLDAYS & ONIONS P.E. Co., Ltd., BIRMINGHAM and LONDON. Established 1650.

ALLDAYS CARS

Three Speeds and Reverse.
Three Powerful Brakes.
Quiet and Smooth Running.

Appreciation:

KING'S LYNN March 7th, 1906.

SIRS,—The 10 b.h.p. Double-cylinder Car which I bought last December has given me the greatest possible satisfaction. It has been driven over 2,000 miles, through rain, mud, slush, and snow, without a single mechanical trouble. My total cost of repairs or replacements up to date has been one [illegible] screw for which I paid 2½d.

ALLDAYS & ONIONS, BIRMINGHAM. Yours truly, J. PLOWRIGHT,

Apply for Catalogue.

ALLDAYS & ONIONS Pneumatic Engineering Co., Ltd., Birmingham.

London Offices & Showrooms: 20, BUCKLERSBURY (Mansion House), E.C. Established 1650.

PHOTOGRAPHS

TOP-LEFT: A 1902 advertisement for the 4 h.p. single cylinder motor car, the "Traveller Voiturette", which featured a water cooled cylinder head. The rear mounted engine drove the rear wheels through a 2 speed gearbox and the car was available as a 2 or 3 seater. A smaller 2 seater "Traveller Voiturette" had been announced as early as 1898, but it is not known whether these actually went into production.

Some early "Traveller" owners sent in testimonials to the company, such as a Mr Craven from Shipston on Stour, who wrote in 1902, "I have much pleasure in stating that I have now had one of your "Traveller" cars for three months and have during that time driven it over 2,000 miles. I find it very simple and easy to drive and manage. It is a reliable car, and a good hill climber." Another proud owner, Mr George Palmer, this time from nearby Sampson Road in Birmingham, wrote, "I am delighted with the car, it is the praise and admiration of all at the hotels and places I put up at. For doctors and commercial travellers it stands out in brilliant prominence as the ideal car."

BOTTOM-LEFT: A 1906 advertisement for the 10 h.p. motor car. In 1904 the company had been awarded a Silver Medal by the Automobile Club for the outstanding performance of their cars during a 6-day reliability trial, and in speed trials carried out in Skegness in 1906 they took both Gold and Silver Medals. Five cars were entered in the hill climbing trials at Shelsley Walsh and all five cars scored wins and even took the Challenge Cup of the President of the Midland Automobile Club.

ILLUSTRATION

RIGHT: Artists impression of the Matchless Works in 1908. The Caretaker's house is shown in the bottom left hand corner with a small open yard and single storey building to the rear. The two storey building to the right of the picture was the Warehouse built in 1906, with the single storey Fitting Shop along side it built in the same year. This probably became the Chassis Erecting Shop. To the right of the two storey Warehouse was an open yard where the three storey building was erected in 1910, but the yard to the extreme right of this picture did not exist, as this land was occupied by houses already built in Walford Road.

BIG BEN, QUEEN'S HONEY AND
MASTERPIECE SOAPS.
MONTHLY PRIZE
COMPETITIONS
HODGSON & SIMPSON LIMITED, WAKEFIELD.
PRIZE MOTOR CAR
FOR
EACH MONTH'S
SUCCESSFUL COMPETITOR.

PHOTOGRAPHS

TOP-LEFT: A 1906 10 h.p. 2 cylinder motor car supplied to Birmingham Corporation Tramway Department for use as a road inspection vehicle.

BOTTOM-LEFT: Lever Brothers who made soap and other products used large numbers of Alldays vehicles and this picture, taken in 1909, shows one of the motor cars offered as a prize each month to retailers of Levers Brothers products who took part in the "Big Ben" and "Queen's Honey" Coupon Competition. The first successful entrant in 1908 was a Bradford family business J. Harrison & Sons, who proudly took delivery of their Alldays car with the registration number AK 1106.

ILLUSTRATION

BELOW-RIGHT: In 1894 Alldays patented an Ambulance based upon a handcart fitted with a stretcher covered by a canvas hood and suitable for use in confined areas such as very narrow streets and alleyways. Just one year later the Ambulance had been redesigned using two bicycles side by side, with their frames welded together by transverse tubes about three feet long. Across these tubes was another frame suitable for carrying a stretcher, covered with a detachable canvas hood and for use in similarly confined spaces, but with the added advantage of having two attendants and more rapid speed to take the patient to hospital. Ambulances like this were adopted by the Birmingham Hospital Saturday Fund who purchased 4 of these machines in 1895 at a cost of £150.

wide cartway, to the top floor of the Warehouse built in 1907. Each floor of the building was 12 feet high and areas of the first floor were designated as a Varnishing Room 18 feet by 40 feet and for Leather Cutting 17 feet by 20 feet. A note written on one of the drawings dated 24th October 1911 indicates that the third floor was not built, leaving it as a three storey building, but that a basement had been added to the building instead. However, no details of the basement were given and no amendments made to the original drawings from the architect Ewan Harper. By 1916 this new building was shown as the Motor Car Erecting Shop on a drawing submitted in June of that year for permission to build an Oil Store in the Fallows Road site in the corner of the open yard occupied by the Wood Store (D) shown on the block plan. The oil store was a brick built building 20 feet by 36 feet and 11 feet high to the flat roof. The need to store large quantities of oil was in itself a sign of how the use of the factory had gradually changed from its early days of cycle production, although cycles were still being produced there for the armed forces. The production of private cars had been drastically reduced to enable the company to meet the demand for commercial vehicles, primarily for the armed forces, but also to meet urgent orders for civilian operators, many of whom were heavily involved with the transport and/or manufacture of war material.

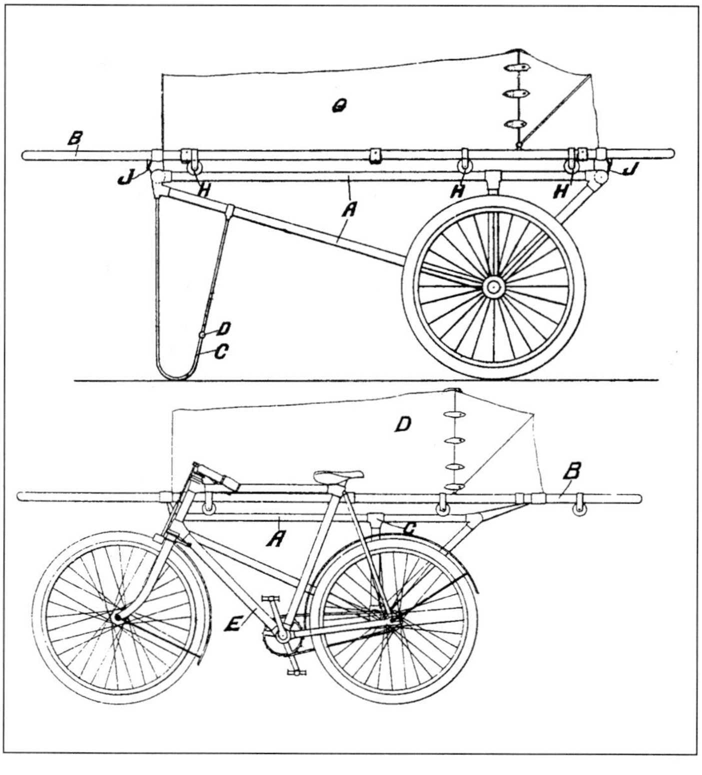

SEE

ALLDAYS

Stand No. 95

STANLEY SHOW.

SEE ALLDAYS MOTOR CYCLE AND METHOD OF FIXING MOTOR TO FRAME.

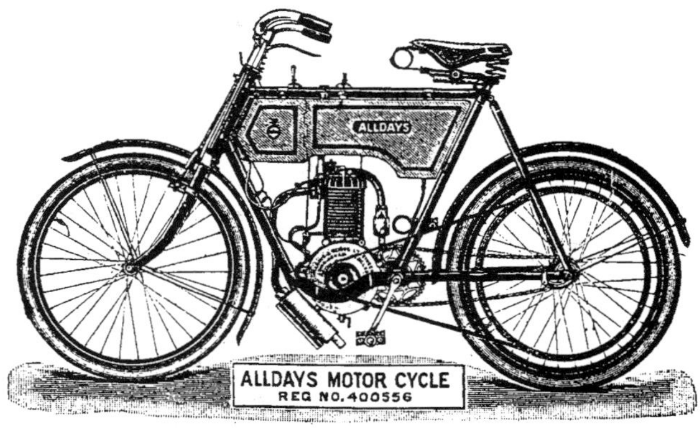

RELIABLE, LIGHT, RIGID, SPEEDY. NO CLIPPING TO TUBES. NO BENT TUBES.

Friends, Old and New, should call and inspect this Fine Exhibition of Cycles, Motor Cycles, Trailers, Carriers, etc. The Largest Exhibit in the Show. Prices and Quality astounding. No better value given anywhere.

ALLDAYS & ONIONS PNEUMATIC ENGINEERING Co. Ltd.

BIRMINGHAM AND LONDON.

(ESTABLISHED 1650.)

MOTORCYCLES

Like many manufacturers of bicycles it became a natural progression for the company to motorise examples of their bicycles when knowledge of the small petrol engine became more widely available at the end of the 19th century. As early as 1898 the company began experiments with a motorised tricycle powered by a proprietary French De Dion Bouton single cylinder engine, but none of these are thought to have been offered for sale.

In 1901 three single cylinder motorcycles were exhibited at the Stanley Show held in London and by 1903 "Alldays-Matchless" 2.25 h.p. single cylinder machines were on sale to the general public. One of these machines was entered in the Autocycle Union trials that year and was awarded a 2nd Class Certificate. These motorcycles cost about £42, much more than the average working class man could afford, but for the middle class man the motorcycle opened up a completely new mode of transport. One step up from a pedal cycle this new form of transport offered a machine no bigger than the cycle, just as easy to store and ride, but without the physical effort to enable long distances to be covered in relative comfort. One owner of a 1904 2.25h.p. machine wrote very enthusiastically to the company to report his journey from Stony Stratford to Bath, 100 miles without stopping, and then on to Taunton, a total of 160 miles, covered at an average speed of 17 m.p.h. without any breakdowns along the way!

At the Birmingham Motor Show held at Bingley Hall, Broad Street, in February 1905 2.5 h.p. and 3.5 h.p. machines were exhibited, both being powered by single cylinder engines.

An entirely new 2.5 h.p. model with an automatic inlet valve was introduced in 1908, the engine being fitted with an Amac carburettor and Bosch magneto. The engine was mounted very low in the frame with petrol fed from a new cylindrical fuel tank. Although the front forks were not sprung the handlebars featured a coil spring device which allowed them to pivot and provide some absorption of road shocks. The braking system featured a front wheel rim brake and a back pedalling belt rim brake on the rear.

However, due to pressure on the works to fulfil large orders for motor cars, production of motor cycles was scaled down drastically between about 1904 and 1908. Then, with various extensions made to the Matchless Works, additional space was made available for the production of motor cycles again, albeit based around one model, the ever popular 3.5 h.p.

The machine shown at the Stanley Show in 1911 was a 3.5 h.p. twin with the cylinders side by side and fitted with a Brown & Barlow carburettor. Transmission was via a Roc two speed gear and belt drive and aluminium footboards were fitted which were sprung from their forward ends. One gentleman, from 'E.K.S.' Cycle Works in Gloucester, wrote to the works in praise of a machine, "Re. motor cycle. We are very pleased indeed with this; it has quite exceeded our expectations. For hill climbing, and reliability it is the acme of perfection. So well are we pleased with it, we should like the Agency for this area."

Six models were shown at the Olympia Cycle Show in 1912, a 2 h.p. single cylinder machine with belt drive; a 3 h.p. twin cylinder machine with belt drive; three 3.5 h.p. single cylinder machines, one fitted with a countershaft chain drive and 3 speeds and two belt driven models, one with a 2 speed Roc patent hub and the other

ILLUSTRATION

This advertisement for the Alldays motor cycle was published in the November 21st 1903 issue of "Cycling" magazine. Like early motor cycles from other makers this design clearly shows how the motor cycle was developed from motorised bicycles.

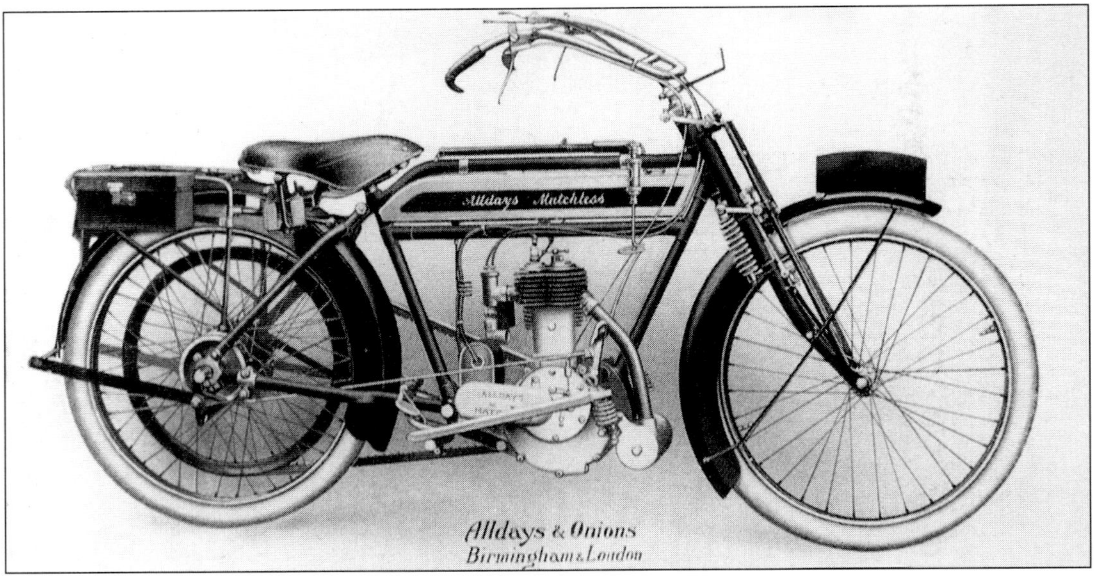
Alldays Matchless
Alldays & Onions
Birmingham & London

Alldays Matchless
Alldays Matchless
Alldays & Onions Ltd
Birmingham & London.

Alldays Matchless
0-Y-22
Alldays Matchless
Alldays & Onions Ltd
Birmingham & London.

PHOTOGRAPHS

LEFT-TOP: A 3½ h.p. "Matchless" motor cycle of 1913 with rubber belt drive and fitted with a two speed hub made under Roc patents. The engine had a bore of 85mm diameter and stroke of 88mm and the complete machine as illustrated cost £50-8s-0d.

LEFT-CENTRE: Another 1913 3½ h.p. "Matchless" machine fitted with chain drive to a 3 speed gearbox and chain drive to the rear wheel. This machine cost £54-12s-0d.

LEFT-BOTTOM: The vee twin engine in this 6/8 h.p. machine had a bore of 85mm diameter and stroke of 88mm. With a 3 speed gearbox and chain drive this top-of-the-range "Matchless" machine cost £73-10s-0d in 1913.

BELOW: A "Matchless" motor cycle with cane sidecar competing in one of the popular trials of this period.

machine with a clutch hub. The sixth machine intended for a sidecar combination outfit, was a 6/8 h.p. vee twin cylinder model with a bore of 85mm diameter, stroke of 88mm, 3 speed gearbox and fully enclosed chain drive. Some idea of the performance and reliability of these machines can be gained from an enthusiastic owner who wrote to the company in July 1912. "I took it to Salisbury Plain, where I was camping for a fortnight, and did a considerable distance on it. Just recently I did four days tour as follows:- Sidmouth (about 175 miles) first day, Sidmouth to Exeter, Teignmouth, Morton Hampstead, across to Dartmoor, to Gairstock, up to Oakhampton, and then to Ilfracombe, second day. Along the north coast of Devonshire, through Lynmouth to Wells, third day. Over 500 miles in all, and the only trouble, a bit of grit or dust under the carburettor needle. I averaged 23-24 m.p.h., excluding stops over a four days run and did about 90 m.p.g. As you will see from my route I had to tackle some good hills, especially at Lynmouth."

Two 3.5 h.p. machines took Gold and Silver Medals in the 1912 Scottish Six Days Open Reliability Trial which was described by one rider as "about the most severe on record". The Gold Medal was gained by a splendid performance covering a total of 1051 miles without losing a single mark.

By 1913 Alldays motorcycles were being entered in the Tourist Trophy races on the Isle of Man usually by private entrants, with mixed fortunes. A machine ridden by W.H.Longton finished in 22nd place, but a second machine ridden by R.E.Surman was forced to retire. These machines were powered by single cylinder 3.5 h.p. engines having a bore of 85mm diameter and stroke of 88mm and featured double foot brakes. Longton's machine also featured an

ALLDAYS
ALLDAYS & ONIONS LTD
BIRMINGHAM & LONDON

ALLDAYS
ALLDAYS & ONIONS LTD
BIRMINGHAM & LONDON.

ALLDAYS
ALLDAYS & ONIONS LTD
BIRMINGHAM & LONDON.

unusual exhaust system whereby the downpipe entered midway along the horizontal silencer which had a small opening at the front end to produce a form of eduction for the exhaust gases.

The 1913 range of "Matchless" motor cycles, all 3.5 h.p. models, started with a belt driven model with pedalling gear at £42 to a belt driven machine with 'Free Engine' and single speed hub at £47-10s-0d, then to a single cylinder chain driven machine at £54-12s-0d and a 6/8 h.p. vee twin with chain drive at £73-10s-0d. A selection of outfits with sidecars was also available, including a tradesman's delivery model known as the "Expressodol" Number 2 at £60-18s-0d.

The Alldays "Matchless" machines for the 1914 season featured an improved and strengthened gearbox, primary gearshaft and improved gear selection mechanism. The 6/8 h.p. vee twin engine was designed for the cylinders to fire at 180°, which was said to give almost perfect torque obtainable from a twin cylinder engine. Both the 6/8 h.p. twin and 3.5 h.p. single cylinder machines had bores of 85mm diameter and strokes of 88mm. Three Alldays machines were entered in the 1914 T.T. races, but all were forced to retire which was a great disappointment for Longton who did particularly well in practice runs.

By 1915 a 2.75 h.p. 292 c.c. patented two stroke model was being produced with a bore of 70mm diameter and stroke of 76mm and featuring three ports, for transfer, inlet and exhaust. This new model was available with a variety of transmissions. A belt drive model with single speed cost £27-10s-0d, 2 speed countershaft drive model £37 and a 2 speed countershaft drive model with a hand operated clutch £40.

PHOTOGRAPHS

LEFT-TOP: By 1915 the Alldays "Matchless" motor cycles were readily identified by the distinctive word "Alldays" cast into the cylinder head. This 3½ h.p. belt driven machine was fitted with a Sturmey-Archer 3 speed hub and cost £52-10s-0d.

LEFT-CENTRE: A "Matchless" 4 h.p. model which had a bore of 86mm diameter and stroke of 95mm. Fitted with a 3 speed gearbox, kick starter and chain drive it cost £58.

LEFT-BOTTOM: The "Matchless" 6/8 h.p. vee twin with 3 speed gearbox cost £76. A 5/6 h.p. vee twin with a bore of 76mm diameter and stroke of 88mm was also in production in 1915. Fitted with a 3 speed gearbox, kick starter and chain drive it cost £74.

BELOW: The caption on this photograph tells its own story. A 1912 "Matchless" machine in service with the British army as a despatch carrier.

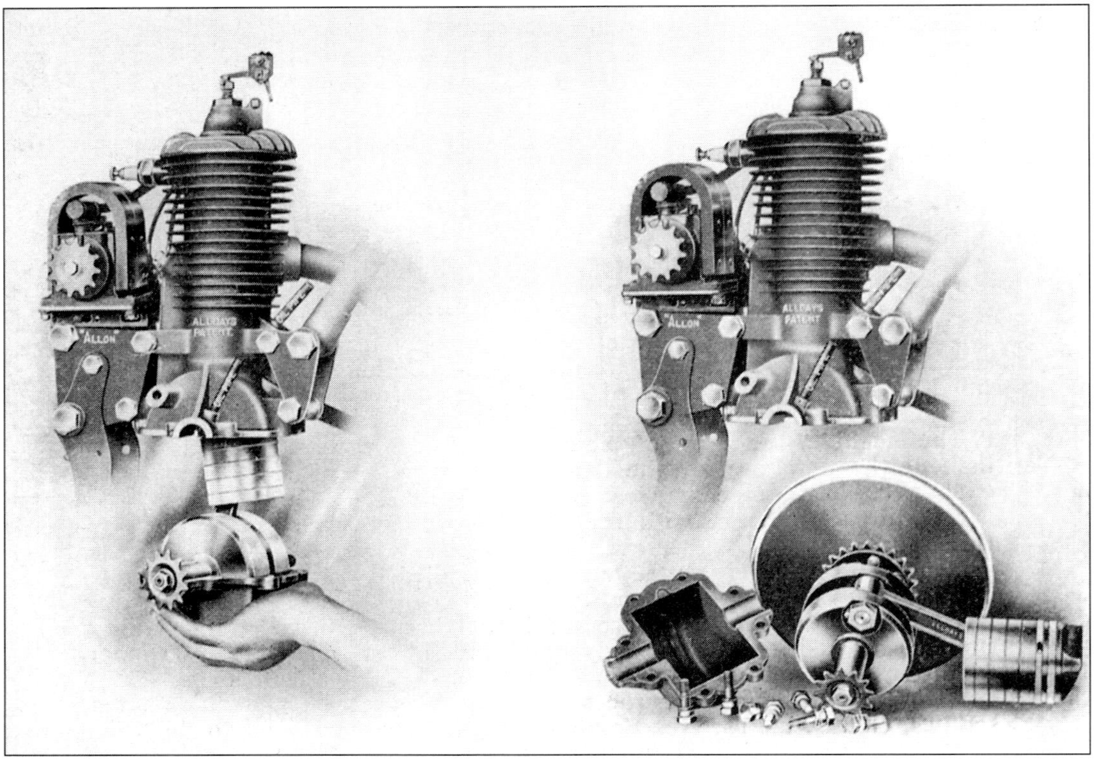

Patent Two-Stroke "Allon" Motor Cycle.

PHOTOGRAPHS

LEFT-TOP: The "Allon" two stroke engine had a bore of 70mm diameter and stroke of 76mm giving it a capacity of 292 c.c. The patented design allowed the crankshaft, bearings, con-rod and piston to be removed, with the engine still in the frame, just by removing the lower half of the crankcase.

LEFT-BOTTOM: A complete "Allon" two stroke motor cycle of 1915 with chain drive from the engine to a countershaft and belt drive to the rear wheel. With single gear it cost £30 and fitted with a 2 speed gear and hand operated clutch it cost £40. A "Matchless" 2 stroke machine was already in production at this time, the engine having a bore of 70mm diameter and stroke of 70mm, giving it a capacity of 269 c.c. A single speed model with belt drive cost £27-10s-0d.

BELOW: A twin cylinder version of the "Allon" two stroke was made with the cylinders side by side and was, more or less, two single cylinder engines bolted together. This 1916 5/6 h.p. machine cost £70 as a solo machine and £85 with the sidecar as shown here. It is believed that only three of these twin cylinder two stroke engines were actually made.

The two stroke engine was always referred to as of 'patented design' in all contemporary Alldays literature. However, the patent, number 7414, was actually taken out by Samuel Upton who lived at 180 Walford Road, Sparkbrook, just a few minutes walk from the Alldays factories. Upton was a Works Manager, possibly at the Great Western Works, but the patent had no other connection with the company, and covered the design feature of being able to remove the main bearings, crankshaft, con-rod and piston without having to remove the engine from the frame. It also covered the design of having a removable plate at the top of the transfer port to allow the port to be hand finished at the point where it entered the cylinder, which would otherwise have been inaccessible.

The company took out several patents associated with the design of motorcycles, the first being taken out in 1915 dealing with the suspension of the rear wheel which was actually used on some 1916 models. The same type of suspension was adopted on a sidecar frame patented in 1917 and this was followed in the following year by a design proposing fittings between a motorcycle and sidecar which allowed rapid attachment and detachment of the sidecar to and from the motorcycle. Whilst this was an ingenious idea it appeared to be insufficiently robust to have withstood the very substantial side loads imposed on such connections between the motor cycle and sidecar during cornering.

The 1915 3.5 h.p. engines featured cylinder heads with the word "Alldays" in large letters across one side of the air fins making them quite distinctive from other machines and a belt driven machine with Sturmey-Archer 3 speed hub and free engine cost £52-10s. The 1915 range included a 4 h.p. single cylinder model with a bore of 86mm diameter and stroke of 95mm fitted with a 3 speed gearbox and

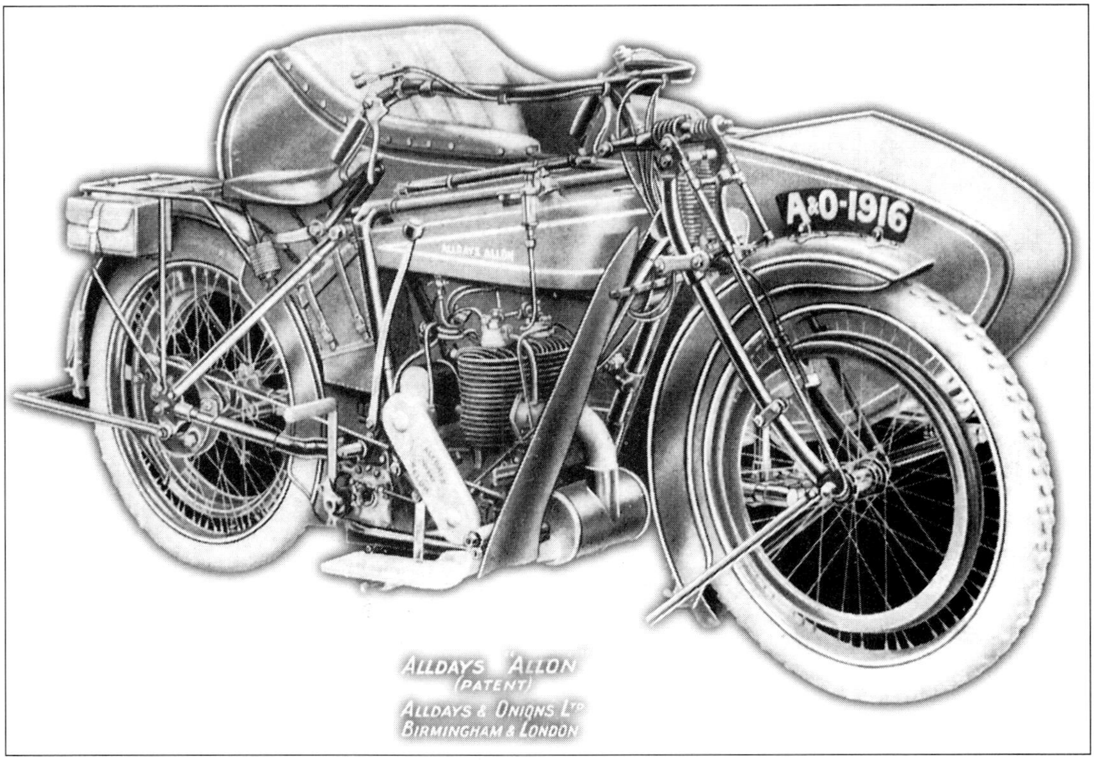

Alldays & Onions Ltd
Birmingham & London.

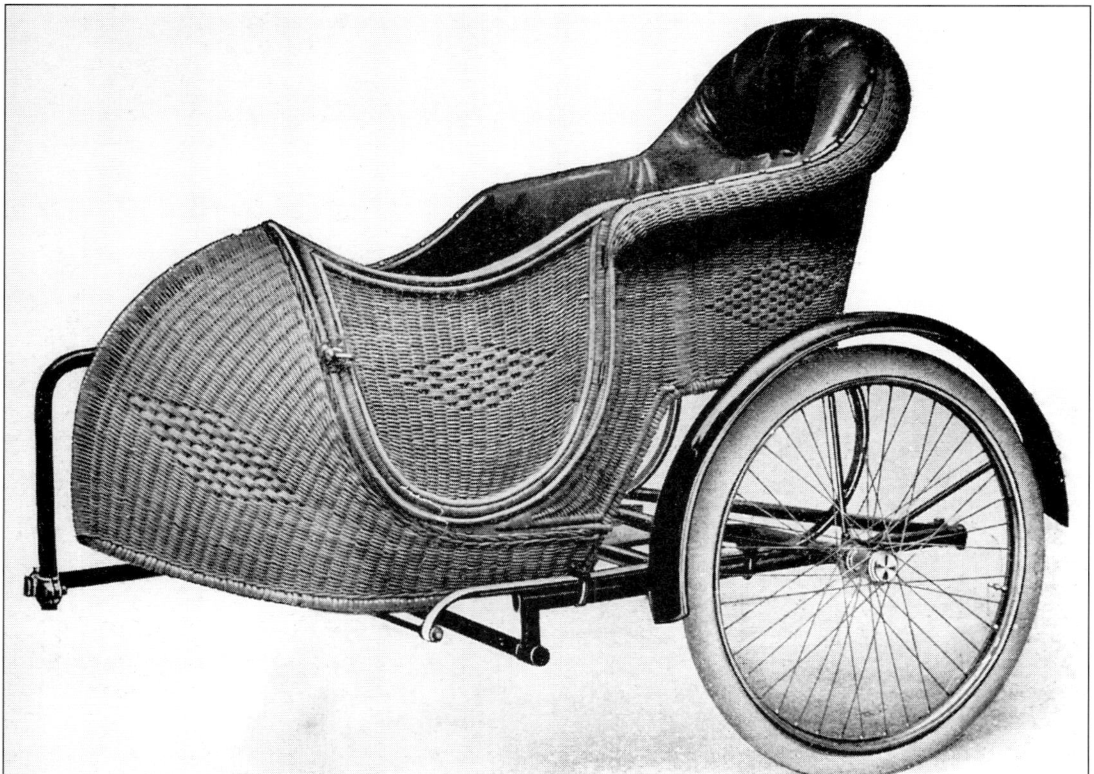

PHOTOGRAPHS

LEFT: Two examples of "Matchless" sidecars from a 1913 catalogue. The sidecar in the top picture looks almost modern in comparison to the basketwork model shown below!

RIGHT-TOP: T.T. rider Mr. W.H. Longton posed beside his Alldays racing machine which finished in 22nd place in the 1913 Isle of Man races.

BELOW: A 1913 3½ h.p. "Matchless" motor cycle with cane sidecar cost £60-18s-0d with two speed Roc hub and £63-0s-0d with an Alldays three speed gearbox.

ALLDAYS & ONIONS LTD
BIRMINGHAM & LONDON

ALLDAYS & ONIONS LTD
BIRMINGHAM & LONDON.

choice of chain or belt drive and a new 5/6 h.p. vee twin with a bore of 76mm diameter, stroke of 88mm, Senspray carburettor, 3 speed gearbox with kick starter and choice of chain or belt drive from the gearbox to the rear wheel. The petrol tank on the twin cylinder machine had a capacity of 2 gallons, compared with 1.5 gallons on the single, both tanks also holding a quart of lubricating oil. With belt drive the machine, which was finished in black enamel with other parts 'heavily plated', cost £72 and with chain drive £74, or for an extra £12, complete with an Alldays sidecar.

The 1916 range of motor cycles was comprised of a single cylinder two stroke "Allon", a de-luxe version of the same machine, a twin cylinder two stroke "Allon" (it is believed that only three of these twin cylinder engines were ever made) and a sidecar combination outfit using the same machine and a single cylinder four stroke "Allon".

A gentleman from Redcar wrote a very complimentary letter to the company, about his 2.75 h.p. two speed two stroke machine in 1915 and upon delivery started a 900 mile touring holiday carrying a lady passenger. No mechanical problems were experienced with the machine during his tour and the motor cycle returned an average fuel consumption of 110 m.p.g. and could reach a top speed of 45 m.p.h. (Other riders reported getting 98 m.p.g.)

The hill climbing ability of the two stroke machine was also praised by another customer from Taunton who reported that he had climbed Peak Hill in Sidmouth which has a gradient of 1 in 7 and is two miles long, in top gear. Salcombe Hill in Sidmouth with a maximum gradient of 1 in 5 was climbed three quarters of the way in top gear and the remainder in second gear. Almost all the owners who wrote in to the company praised the two stroke machines for their hill climbing ability in top gear and with never a hill being found to beat the machine.

PHOTOGRAPHS

LEFT-TOP: A 1913 "Matchless" motor cycle with a tradesman's delivery side car. With a 3½ h.p. machine the outfit cost £65, with a 4 h.p. machine £67-10s-0d or with a 5/6 h.p. machine £84-10s-0d.

LEFT-BOTTOM: A 3½ h.p. "Matchless" motor cycle fitted with 3 speed belt drive and coupled to a company made side car cost £65-10s-0d.

BELOW: Another 3½ h.p. "Matchless" motor cycle with side car cost £63 with a 2 speed Roc hub, or £65-2s-0d with 3 speed Alldays gearbox.

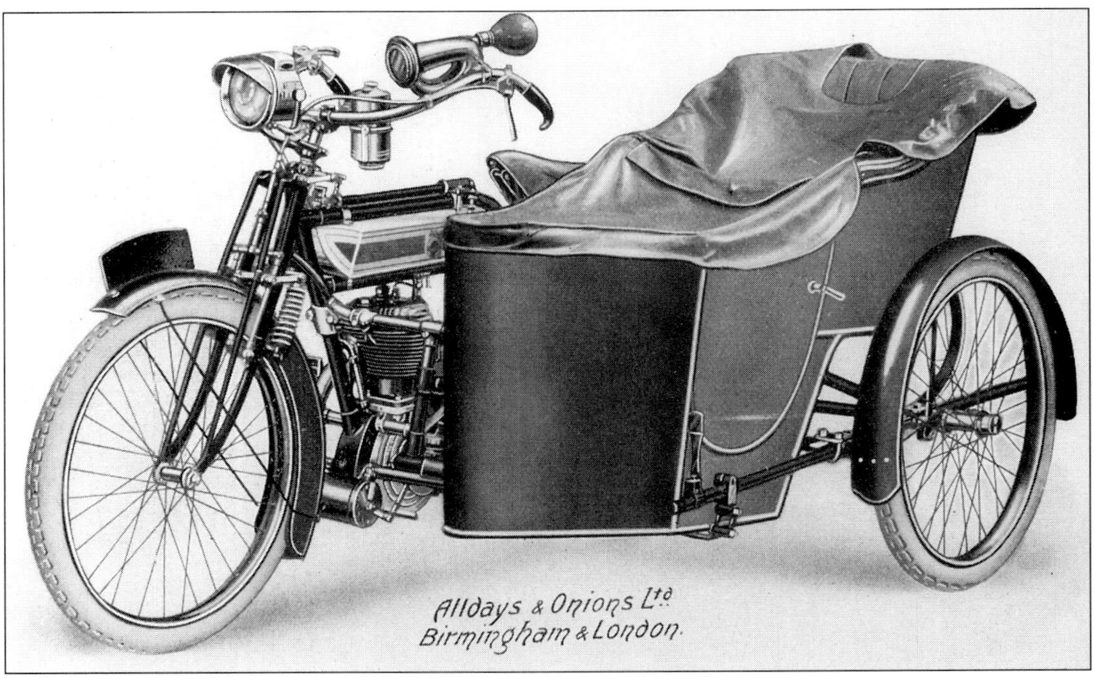

IO·23

A&O
ALLDAYS "ALLON"
ALLDAYS "ALLON" Model de Luxe
"THE BEST OF ALL MOTOR CYCLES."

PHOTOGRAPHS

LEFT-TOP: A 1916 picture showing an "Allon" two stroke machine described as "The Ideal Lady's Mount". With countershaft single gear it cost £36, or with two speed gear £42.

LEFT-BOTTOM: The 1920 "Allon" de-luxe two stroke machine, "Built by Engineers". Fitted with a two speed gearbox, hand operated clutch and kick starter the price was £85.

ILLUSTRATION

RIGHT: An Alldays advertisement for their "Allon" motor cycle, about 1920.

ALLDAYS 6/8 H.P. ENGINE.

A motor cycle dealer from Newcastle on Tyne wrote, "We find the demand is very much greater than the supply, and the machines generally are giving satifaction to our customers. Please let me know when you can despatch some more." A customer from the same town wrote, "It is the most delightful and simple solo machine that I have ever ridden and never gives the slightest trouble."

By 1920 the "Allon" two stroke machine with countershaft drive, two speed gear, kick starter and hand operated clutch, but still retaining a belt drive to the rear wheel, cost £85. The prices for spare parts makes interesting reading, with an engine cylinder costing £3-10s, a new flywheel £1-10s, a chain sprocket 12s-6d, chain sprocket with cork inserts which formed part of the clutch 15s-0d, change speed lever and quadrant 8s-0d, front forks complete £4, handlebars complete with grips 18s-6d, frame, less forks and stays £5-10s, and rear wheel complete with hub and belt rim £4. A new magneto cost £6-10s, a new carburettor complete with control cable £2-10s, and if a new fuel tank was required it cost just £4.

The two stroke machines were produced in limited numbers up to about 1923/4 when the company ran into financial problems. Motor-cycle production ceased for a short time during which the company was reformed into New Alldays & Onions Ltd., and then small numbers of 680 c.c. vee twin JAP engined machines with twin exhaust silencers were built, selling at £75, together with Dorman-Bradshaw single cylinder oil cooled models, but the two stroke machines were dropped from the company's product range.

PHOTOGRAPHS

LEFT-TOP: A close up view of a 1913 6/8 h.p. vee twin engine which featured a crankshaft running on ball bearings.

LEFT-BOTTOM: An impressive "Matchless" 6/8 h.p. three speed motor cycle with side car which cost £88-10s-0d in 1915. An owner of a 6/8 h.p. combination wrote a letter of appreciation to the company in 1914 stating that it had given every satisfaction. Having ridden it over 3,500 miles in three months he found that its hill climbing capabilities "leave nothing to be desired", and the petrol consumption was averaging about 70 m.p.g.

RIGHT: A sectioned view of an Alldays 3 speed motor cycle gearbox.

503

ALLDAYS & ONIONS LTD
BIRMINGHAM & LONDON.

TROLLEY BUSES

The first trolley bus, or 'trackless trolley' as it was originally called, appeared in Britain in 1909 operating on a short length of experimental line erected in Hendon, London. This line and the trolley operating on it had been designed by The Railless Electric Traction Company Limited of 56 Moorgate Street, London, a company which had no actual manufacturing facilities of its own, necessitating the use of contractors to build line systems and vehicles to their designs.

Two of the first municipal undertakings to introduce 'trackless trolleys' were Leeds and Bradford who adopted the Railless Electric Traction system in 1911. The first four vehicles to be put into service in Leeds during 1911 and 1912 had chassis built by Alldays & Onions and bodies built by Hurst Nelson Co.Ltd., (tramcar builders) of Motherwell in Scotland.

Just how Alldays & Onions became involved in this work is not known, but the company was already building conventional petrol powered passenger and commercial vehicles by this time and was probably awarded this contract on the strength of its reputation for quality in the general field of engineering.

The trolley bus chassis were based upon a channel section steel chassis frame mounted very high by semi-elliptic steel road springs, the chassis having a wheelbase of 13 feet and frame width of 4ft 4ins. The front steering axle was of conventional design and had a track of 6ft 4ins. Artillery type wheels were fitted front and rear, with twins at the rear, and were fitted with solid rubber tyres supplied by the Midland Rubber Co.Ltd. The front wheels were 32" diameter and the rear driving wheels were 36" in diameter.

Two 20 h.p. Siemens electric traction motors were fitted midway along and beneath the chassis frame (hence the very high chassis frame which required four fairly high steps to climb into the passenger compartment) driving via a short shaft into a transverse counter shaft and then by chains to the rear wheels.

Two foot operated brakes were provided, one operating on the counter shaft and the second operating internally expanding brake shoes at each rear wheel. A handbrake lever was also provided for parking which also operated through the rear wheel brakes, but this was said to have been inadequate to hold the vehicle on a steep incline.

The trolley bus bodies had seating for 28 passengers in vehicles which had an overall length of 20ft 1in and were 6ft 8ins wide and were 10ft high, excluding the poles mounted on the roof used to collect electricity from the overhead wires. Each trolley bus cost £725 and had an unladen weight of 5 tons 2 cwts. Despite their heavy unladen weight the Bradford trolley buses, which covered some 55 miles of routes, had running costs of 6.6 pence per mile resulting in both of the neighbouring municipalities making a slight profit from their operation.

The Railless/Alldays trolley buses were also supplied to Dundee and Rotherham in 1912 and the last of the Leeds vehicles, fleet number 504, was withdrawn from service in 1924. Alldays involvement in the building of trolley bus chassis is thought to have been short lived as the Railless company used several manufacturers to build vehicles to their designs before the company itself was taken over by Short Brothers Ltd., in 1918.

PHOTOGRAPHS

TOP: The Leeds "Railless" Trolley number 503 featured a chassis built by Alldays & Onions. The chassis frame was of simple construction, but quite substantial proportions which, together with the heavy duty solid tyred wheels and two heavy electric traction motors, one of which can be seen slung from beneath the chassis frame, helped to give this single deck trolley bus an unladen weight of over 5 tons. The two traction motors drove into individual gearboxes and then via chains to each rear wheel.

Some company records for 1911 show that a total of six trolley bus chassis were built for The Railless Electric Traction Company. The first four were referred to simply as Tractor 'A' Type and the final two as Tractor 'B' type, the difference being that 'A' types were powered by two electric motors and 'B' types by three electric motors. Some of the Leeds trolleys were later converted to front wheel drive and one is left wondering if the two 'B' type chassis were equipped with four wheel drive! The six chassis were numbered 3924, 3926, 3928, 3930, 3932 and 3934, all being delivered during June and July 1911.

Interestingly, there is also reference to a Tramcar having been made in October 1912 for a company called Dunkin & Co., with chassis number 4684 and powered by a 16/20 h.p. petrol engine!!

BOTTOM: With the company engaged in the production of many diverse products it was not surprising that they also built some marine engines, like this 14 h.p. four cylinder unit mounted on its own sub-frame and complete with forward and reverse gear mechanism and even an instrument panel ready for installation.

EXPANDING THE MOTOR CAR BUSINESS

Alldays adopted a simple way to describe their cars in period catalogues as Car No. 1, through to Car No. 9 etc., and in 1905 their Car No. 2, a 2 seater model could be had with a 7 h.p. engine for £170 or with a 8 h.p. engine for £180. Car No. 3, likewise, was offered at £180 or £190 fitted with a 4 seater body. Car No. 4 fitted with a 8 h.p. engine was available with a 4 seater body, the nearside front seat of which swung out to give access to the rear seats and could be detached to make a 3 seater and priced at £197. Car No. 5 was a 2 seater Landaulette with folding hood, "highly recommended for professional gentlemen" and fitted with a 8 h.p. engine all for just £250. The 8 h.p. Car No. 6 priced at £210 was actually a van to carry 5 cwts, but the top of the van body could be removed enabling the vehicle to be converted to a 4 seater motor car. Car No. 7 was also a 8 h.p. model fitted with a 4 seater body complete with folding full length Cape hood for £213-10s-0d and cars Nos. 8 and 9 were 4 seater models specially designed for tropical climates enabling the 8 h.p. engines to be started on petrol and then run on paraffin. With a full length detachable canopy, but open sides and back, it cost £215 and as an open car with cane sided seats £205. All these models were available in Alldays standard colour, dark green, with primrose lining and upholstery to match.

By 1906 the Matchless Works was producing motor cars of 7 h.p., 8 h.p., 10 h.p., and 16 h.p. with prices ranging from £165 to £375 and small vans with a capacity of 5 cwts and 10 cwts were being offered powered by single cylinder 8 h.p. and twin cylinder 10 h.p. engines respectively, priced at £190 and £220. With the rapid acceptance of the motor car the general public were now more interested in motorised forms of personal transport than pedal powered and the production of cycles went into decline. In 1907 a new 14 h.p. motor car was introduced powered by a 4 cylinder engine with a bore of 82mm diameter and stroke of 102mm based on a 8ft 6ins wheelbase chassis frame with a track of 4 feet and priced at £325. The motor car range also included 10 h.p. and 20 h.p. models and the commercial vehicle range now included a 10 h.p. 2 cylinder 10 cwts van with pneumatic tyres for £200, a 15 cwts van powered by the same engine, but fitted with solid tyres, for £210 and a new 30 cwts van powered by a 4 cylinder 20 h.p. engine and fitted with solid tyres priced at £378. At the Olympia Motor Show that year the company also exhibited a small and very unusual motorised railway inspection car powered by a 8 h.p. single cylinder engine and fitted with a 2 speed gearbox. These railway inspection cars were to become a popular addition to many of the railway companies operating in British colonies all over the world, sometimes used alone and sometimes coupled to small passenger carrying open carriages. Two rail cars delivered to Chile in 1909 were reported to be giving excellent service one year later over tracks which were rough and poorly maintained and with gradients of 3%. They had been used at average speeds of 18 miles per hour and, where the tracks permitted, at speeds up to 24 miles per hour. Rail cars had also been exported to New Zealand and Australia, the latter vehicle, a four seater, being put into service in May 1908 on a 24" gauge line connected with the

PHOTOGRAPHS

TOP: In 1908 this 18 seater Chara-banc was built for service in the Far East. The 20 h.p. four cylinder engine drove through a 3 speed gearbox to a transverse cross-shaft then by open chains to the rear wheels. The seating looks very uncomfortable by modern day standards, but the vehicle was only capable of achieving an average speed of about 16 m.p.h. and the side curtains provided some protection from the weather. Like all Alldays motor vehicles this Chara-banc had a simple ladder type chassis frame with parallel side members devoid of any taper to the front or shaping over the rear axle.

BOTTOM: Front view of a "Colonial Car" with a good view of the fully elliptic road springs which gave the chassis enough height to be able to wade through water up to 2 feet deep without getting the driver's feet wet. With members of the Alldays family travelling around the world getting sales they had first hand knowledge of what sort of conditions their vehicles were expected to operate in and could specify vehicles to suit the widely varying conditions.

PHOTOGRAPHS

LEFT: The top picture shows a 1909 Coupe, Model No.55 in 10/12 h.p. form which cost £316-1s-0d, or Model No.56 in 14/18 h.p. form at £383-5s-0d.The 14/18 h.p. model was based on a 8ft 6ins wheelbase chassis with front track of 4ft 3ins and the car had an overall length of 12 feet.

The bottom picture shows a 1909 Single Landaulette with front extension and folding glass screen, Model No.31 in 14/18 h.p. form at £430-10s-0d, or Model No.33 in 20/25 h.p. form at £483. Built on a 9ft 6ins wheelbase chassis with a front track of 4ft 6ins, this model was available upholstered in French Cord cloth or in best quality leather.

BELOW: The Alldays 4 speed and reverse gearbox which featured a direct drive 3rd gear and indirect drive 4th gear, the 4th gear having a slightly higher ratio than 2nd gear. Hoffmann ball bearings were fitted throughout the gearbox and this picture clearly shows the contracting band brake mechanism to the right hand side of the gearbox.

From chassis numbers 4360D3 and 4380B5 the 4 speed gearboxes became available with direct drive in 4th gear, giving road speeds of approximately 7 m.p.h., 16.5 m.p.h., 26 m.p.h. and 32 m.p.h. at 1,000 r.p.m. engine speed. the change in specification taking place about February 1912.

construction of the Burrinjuck Dam in New South Wales and is believed to have worked on the 25 mile long line until completion of the dam in 1928. In New Zealand an 8 seater rail car ran between Railton and Sheffield based upon a 5 feet wheelbase chassis and the vehicle was about 10 feet long overall.

In August 1907 a 10 h.p. type 2 motor car, chassis number 1792BP, engine number 2470, was delivered to Angus, Sanderson & Co. and in April 1908 another 10 h.p. type 21 motor car, chassis number 2020BU, engine number 2810, was delivered to Sir W. Angus Sanderson & Co. (This company itself would enter into motor car manufacturing and ran into serious financial difficulties after the First World War when it tried to mass produce a motor car in conjunction with E.G.Wrigley & Co.Ltd., and the Tylor engine company.)

Although some motor car manufacturers were experiencing a slump in sales Alldays decided to expand their motor business further and acquired the Enfield Autocar Company Limited in April 1908 for an undisclosed sum. This Redditch based company had been making cycles, motor cycles and motor cars under one company called the Enfield Cycle Company Limited, but in 1906 they decided that the motor car side of their business had to stand on its own feet and a separate company had been formed known as the Enfield Autocar Co. Ltd. Although substantial orders had been forthcoming they had experienced difficulties in finding skilled labour, machinery and materials and by September 1907 the new company had lost £19,264. In April 1908 it was decided to put the motor car company into voluntary liquidation.

An auction was held at the Grand Hotel in Birmingham on the 6th April, but the reserve price of £18,000 was not reached and they

ALLDAYS FOUR-SPEED GEAR BOX AND TRANSMISSION.

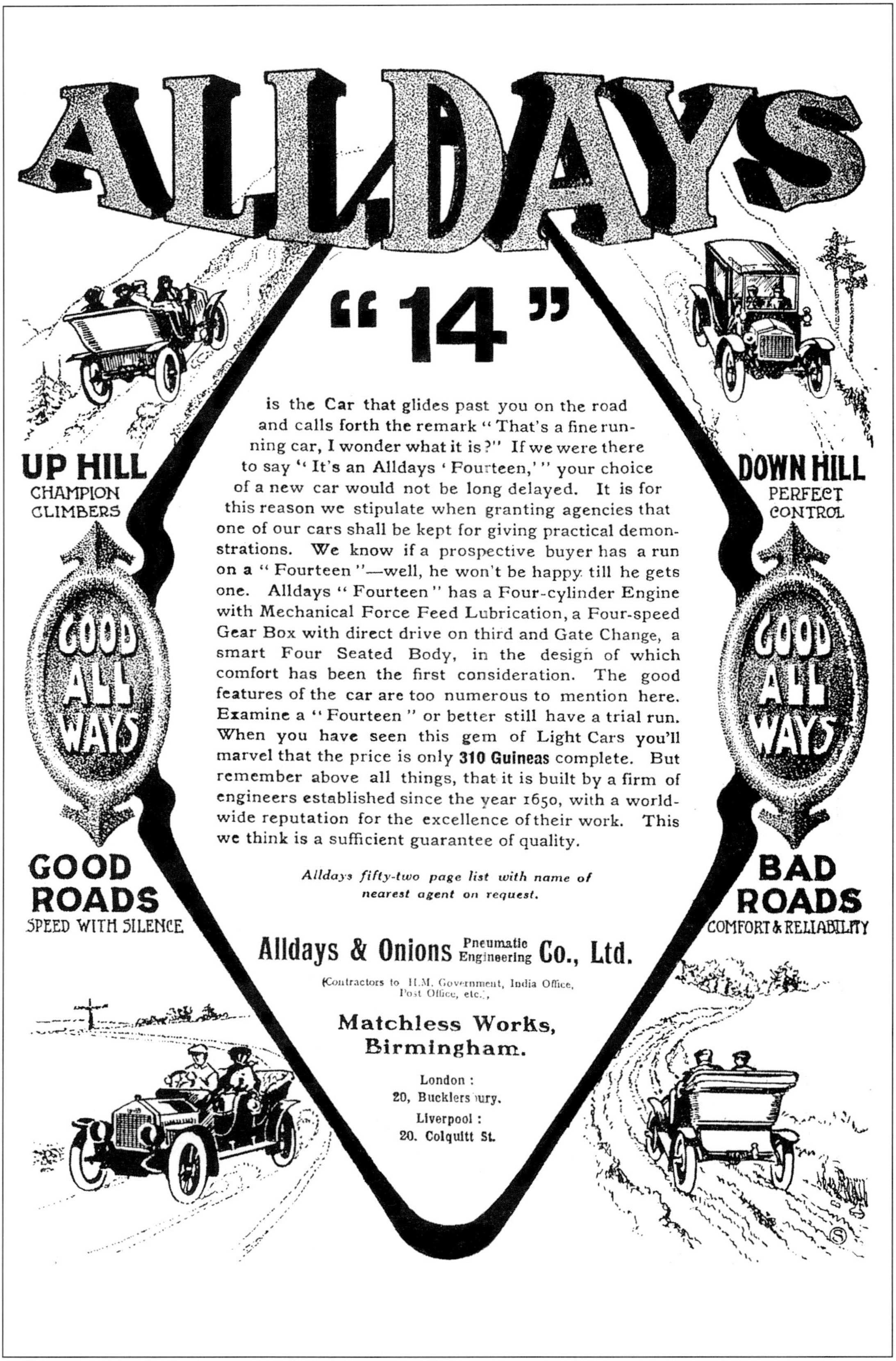
ALLDAYS
"14"
is the Car that glides past you on the road and calls forth the remark "That's a fine running car, I wonder what it is?" If we were there to say "It's an Alldays 'Fourteen,'" your choice of a new car would not be long delayed. It is for this reason we stipulate when granting agencies that one of our cars shall be kept for giving practical demonstrations. We know if a prospective buyer has a run on a "Fourteen"—well, he won't be happy till he gets one. Alldays "Fourteen" has a Four-cylinder Engine with Mechanical Force Feed Lubrication, a Four-speed Gear Box with direct drive on third and Gate Change, a smart Four Seated Body, in the design of which comfort has been the first consideration. The good features of the car are too numerous to mention here. Examine a "Fourteen" or better still have a trial run. When you have seen this gem of Light Cars you'll marvel that the price is only 310 Guineas complete. But remember above all things, that it is built by a firm of engineers established since the year 1650, with a world-wide reputation for the excellence of their work. This we think is a sufficient guarantee of quality.
Alldays fifty-two page list with name of nearest agent on request.
Alldays & Onions Pneumatic Engineering Co., Ltd.
(Contractors to H.M. Government, India Office, Post Office, etc.),
Matchless Works, Birmingham.
London: 20, Bucklersbury.
Liverpool: 20. Colquitt St.
UP HILL
CHAMPION CLIMBERS
DOWN HILL
PERFECT CONTROL
GOOD ALL WAYS
GOOD ALL WAYS
GOOD ROADS
SPEED WITH SILENCE
BAD ROADS
COMFORT & RELIABILITY

withdrew from the auction. A few days later Alldays bought the company and ran it still under the name of the Enfield Autocar Company Limited, but from its own Matchless Works in Birmingham. (A letter written by William Allday in January 1910 states, "Some 18 months ago we bought the Enfield Autocar Co's Works, at Redditch, and unfortunately we have not been able to do anything with the Works".) Some of the motor cars in the 1908 range included a 10 h.p. with a 2 cylinder engine having a bore of 95mm diameter and stroke of 114mm, fitted with a three speed gearbox and based on a 6ft 6ins wheelbase chassis frame, priced at £200 for a chassis or £220 for a complete 2 seater car; a 14 h.p. powered by a 4 cylinder engine with a bore of 86mm diameter and stroke of 102mm and based on a 8ft 6ins wheelbase chassis priced at £262 in chassis form or £325 with a four seater body. A 20 h.p. car powered by a 4 cylinder engine with a bore of 95mm diameter and stroke of 114mm and based on a 9ft 6ins wheelbase chassis cost £367 in chassis form and £451 with a five seater body.

Between 1908 and 1910 five 20 h.p. 24 seater charabancs were built at the works for Alldays London dealer McIlwraith, McEachern & Co., the first three, chassis numbers 2142CD, 2144CD and 2148CD, were based on chassis with a front track of 4ft 6ins and the last two, chassis numbers 2196CJ and 2198CJ, based on chassis with a front track of 4ft 8ins. All five vehicles were fitted with chain drive to the rear wheels which had solid tyres front and rear. The company announced that it had made a net profit of £13,540 for the year 1907/08.

Having quite a low output the company was in an excellent position to build vehicles to customers' own specifications. It is not surprising, therefore, to find 10 h.p. motor car chassis being built for Tomlinson & Sons, the London dealer, with chain drive on a 5ft 4ins short wheelbase frame and with a 4ft 8ins front track. Six of these were built in 1907, together with two more chain driven chassis with a front track of 3ft 10½ins. In the following year four more special chassis were built for C.A.Moody – these were also 10 h.p. models, but with 6ft 6ins wheelbase chassis and 3ft 10½ins front track.

In 1909 Alldays entered the motor cab business producing two 4 passenger models, one powered by a 2 cylinder 10 h.p. engine, priced

ILLUSTRATION

LEFT: Advertisement for the Alldays '14' motor car from the April 24th 1909 issue of "The Autocar" magazine. Even this advert claims "that it is built by a firm of engineers established since the year 1650". The price of the motor car was 310 guineas, or £325-10s-0d.

PHOTOGRAPH

BELOW: A 1909 Commercial Travellers Car, Model No.12 powered by a 10/12 h.p. 2 cylinder engine. The rear box like compartment was fitted with two doors and was used for carrying samples or delivering goods to retailers. Additional items could be carried on top of the box which was equipped with rails for this purpose. This motor car cost £231.

ALLDAYS &ONIONS LTD
ENGINEERS
BIRMINGHAM & LONDON.
0 4694
Alldays & Onions Ltd
Birmingham & London.

at £226 plus tyres, and the other model being powered by a 4 cylinder 14 h.p. engine, priced at £268 plus tyres. Both of these engines had a bore of 3.75 inches diameter and stroke of 4.5 inches and were fitted with four speed gearboxes. Alldays quoted prices, plus tyres, because there was such a large selection of different types and makes available it was left up to the customer to choose and then add the cost onto the price of the complete motor car. The range of vans was now shown as 15 cwts, 30 cwts and 40 cwts models, the first two being powered by 2 cylinder vertical engines of 10 h.p. and 14 h.p. respectively priced at £231 and £290 plus tyres and the 40 cwts model was powered by a 4 cylinder 20 h.p. engine and cost £332 plus tyres. All these vehicles featured a 3 speed gearbox.

The little motor railcars now included models of 8 h.p., 10 h.p. and 14 h.p. to seat 4 to 5 passengers and a 9 to 12 passenger model was also listed which could be driven from either end.

The motor car range comprised two 10 h.p. models powered by 2 cylinder engines with a bore of 95mm diameter and stroke of 114mm on a 7ft 6ins wheelbase chassis at £199 in chassis form or £221 as a complete car and on a 8ft 0in wheelbase chassis at £220 in chassis form and £262 as a complete car. A 14 h.p. model was available powered by a 4 cylinder engine with a bore of 86mm diameter and stroke of 108mm and based on a 8ft 6ins wheelbase chassis which cost £262 in chassis form and £325 as a complete car. The top of the range model was of 20 h.p. car powered by a 4 cylinder engine with a bore of 95mm diameter and stroke of 114mm, based on a 9ft 6ins wheelbase chassis which cost £315 in chassis form and £378 as a complete car. All models in the motor car range were fitted with 4 speed gearboxes.

For the 1910 season a completely new 6 cylinder engined car was introduced designated the Alldays 30/35 h.p. model, the engine having a bore of 95mm diameter and stroke of 114mm with the car based on a 10ft 6ins wheelbase chassis. The rest of the car range was re-designated 10/12 h.p., 14/18 h.p. and 20/25 h.p. models, the only changes made being that the cheapest 2 cylinder model had a 3 speed gearbox and the 14 h.p. model was based on a 9ft 0in wheelbase chassis.

By this time special "Colonial" cars were being offered with the same choice of engines, but fitted with full elliptic road springs and 30" diameter wheels giving a ground clearance of 14" and could wade through 2 feet of water. The drive to the rear wheels on these models was by two side chains and the cars were priced at £288-15s-0d, £341-5s-0d and £393-15s-0d for 10/12 h.p., 14/18 h.p. and 20/25 h.p. respectively. A 14 h.p. "Colonial" car, chassis number 2100BY with engine number 2938, was exported to the Australian agent Dalgety & Co. in May 1908, the engine being fitted with a Brown & Barlow carburettor and E.I.C. ignition coil made in nearby Sampson Road North. Another "Colonial" car, chassis number 2098BY, was delivered to a customer in July of the same year, the body being fitted with a double Cape Cart hood with folding glass screen, painted red with brass mounts and the 14 h.p. engine, number 2924, fitted with a Watson ignition coil.

In New Zealand about this time one Alldays car was being used in something of a pioneering role when Bert Mitchell and Ivan Wintringham, who were agents for Alldays & Onions, drove a 8-10 h.p. car to Tophouse in an area with no roads, driving through bush and fording rivers along the Wairau Valley.

PHOTOGRAPHS

TOP: Model No.35 14/18 h.p. Taxicab, or Commercial Landaulette. Built to conform to Scotland Yard Regulations this particular model gave little in the way of protection for the poor driver who had to endure all weather conditions, whilst the passengers were seated in the rear compartment, the hood of which could be lowered in fine weather. Another version of this Taxicab was available fitted with a roof extension over the driver, but still with no windscreen or doors to the driving compartment. This model cost £409-10s-0d.

BOTTOM: This unusual looking vehicle is a forward control Alldays commercial of about 15/20 cwts load capacity fitted with canvas tilt van body, solid tyres front and rear and only partially enclosed chain drive to the rear wheels. According to the sales brochure it was intended for use by wholesale grocers and provision merchants.

ENFIELD AUTOCARS

Some alternative styles of Bodywork

FOR WHICH WE SHALL BE PLEASED TO QUOTE ON APPLICATION.

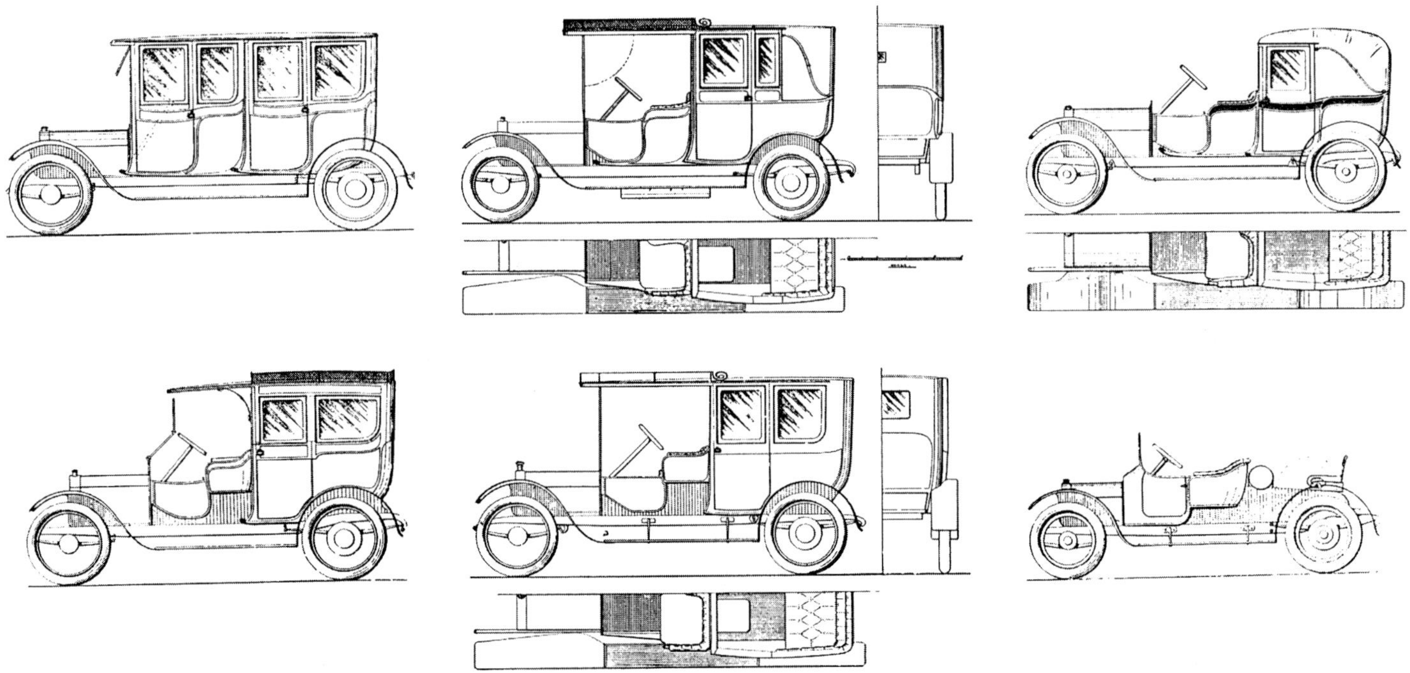

TRADE MARK. MADE LIKE A GUN.

Letters of appreciation from customers make interesting reading, like a gentleman from Bangalore in India who wrote in to the company to tell them that his Alldays car had made a record ascent of the Ootacamund Ghaut from Mettapollium covering a distance of 32 miles with a rise of 8,000 feet, carrying three people and their luggage, in 2½ hours. A London customer reported more fully about his 14 h.p. car which covered over 8.000 miles in the first year, used 382 gallons of petrol at a cost of £20-9s-9d, with oil and grease costing £3-17s-8d, repairs totalling£2-2s-6d, sparking plugs and spare springs cost £1-7s-3d and the greatest expense of all, tyres at a cost of £20-13s-1d.

A Captain from the School of Musketry in Kent had just covered his first 1,000 miles in his 10 h.p. 2 seater with total running costs of £2-19s-4d. The little car was averaging 27 miles per gallon and averaged 25 miles per hour. As regards "going capabilities" he only needed to change twice from top gear to second in a journey from Croydon to Portsmouth and, "she simply flew up the Devil's Punch Bowl at Hindhead".

The van range had been changed to a 10 cwts, 15 cwts, 20 cwts and 30 cwts powered by a 10/12 h.p., 14/18 h.p. and 20/24 h.p. in both the 20 cwts and 30 cwts. However, the van prices had now risen to £256-15s-0d, £330-15s-0d, £420-0s-0d and £446-5s-0d respectively, plus tyres which ranged from £12-5s-0d to £21-5s-0d per set.

The company was offering two different ranges of cars by this time, one under its own name of Alldays and a second range of cars under the Enfield Autocar name.

The Enfield range for 1911 included a 10/12 h.p. model using the Alldays 2 cylinder engine, 3 speed gearbox and 7ft 6ins wheelbase chassis, a 16 h.p. model which used the Alldays 4 cylinder 14/18 h.p.

ILLUSTRATION

LEFT: A page from a 1910 Enfield Autocars sales catalogue illustrating alternative body styles available on Enfield chassis. A similar illustration appears in the Alldays catalogue of the same period showing how the two ranges of cars, which differed mechanically in many ways, shared bodywork applicable to all the motor cars then being built at the Fallows Road Matchless Works.

PHOTOGRAPH

BELOW: The 1910 Enfield 18/20 h.p. Model 181, priced at £367-10s-0d complete with leather upholstery, or at £304-10s in chassis form. This model featured a four cylinder engine with a bore of 95mm diameter and stroke of 115mm with high tension magneto, leather cone clutch and four speed gearbox. It was based on a 9ft 3ins wheelbase chassis which had an overall length of 12ft 6ins and overall width of 5ft 3ins.

A.D. 1911. Nov. 13. No. 25,189.
ALLDAYS & ONIONS PNEUMATIC ENGINEERING CO. & another's COMPLETE SPECIFICATION. [1 SHEET]

[This Drawing is a reproduction of the Original on a reduced scale]

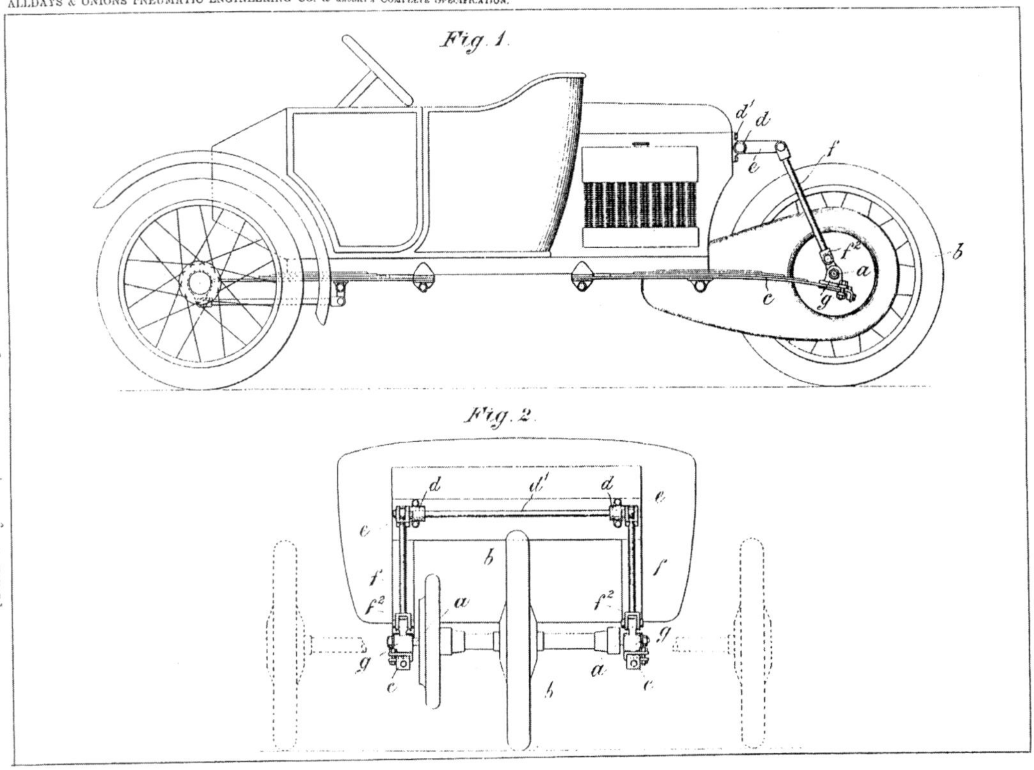

A.D. 1912. Jan. 20. No. 1633.
ALLDAY & another's COMPLETE SPECIFICATION. [1 SHEET]

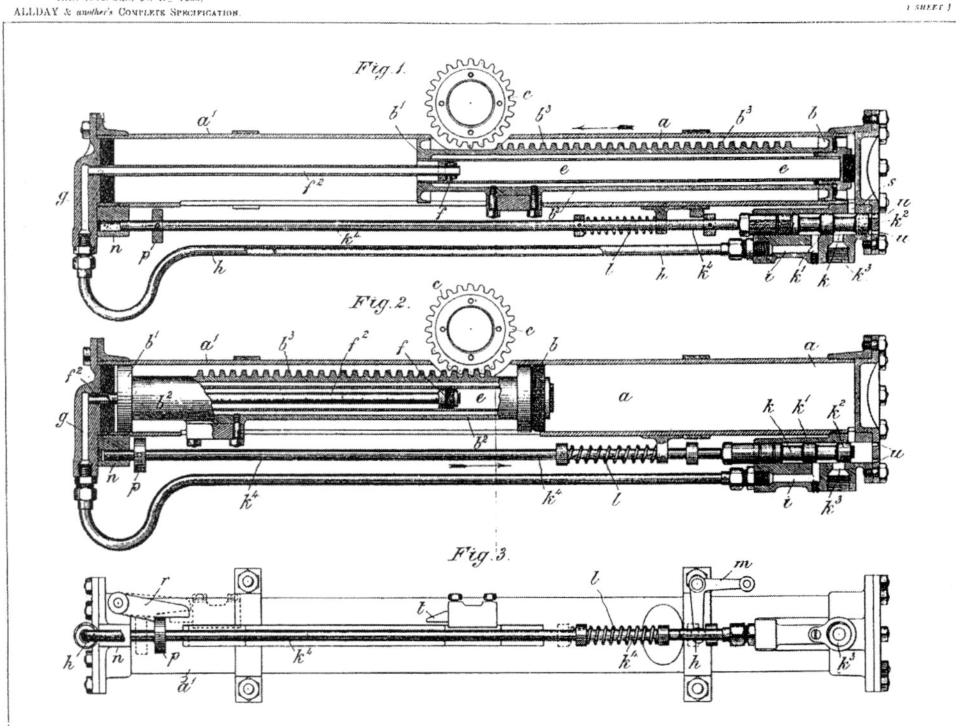

Malby & Sons, Photo Litho.

ILLUSTRATIONS

LEFT-TOP: This diagram illustrates a three wheeled motor car designed in 1911 by Charles Edward Simms, an engineer with the company, and the subject of Patent Number 25189. The two front steering wheels were mounted on a conventional axle bed and supported by cantilevered laminated road springs. The engine was mounted to the rear of the seats, the drive to the single rear wheel(b) being transmitted by an enclosed chain.

The principle feature of this design was the positive location of the rear wheel axle(a) which was supported by two widely spaced, cantilevered laminated road springs(c). At the point where the driven axle is supported by the springs two arms(f) extend forwards and upwards and are connected to levers(e) which are fixed to the horizontal shaft(d) mounted across the rear of the motor car in bearings(d). This method of locating the single rear wheel was designed to prevent excessive canting over of the wheel on rough road surfaces and during cornering and helped to prevent torsional twisting of the rear road springs and their mountings to the short rear axle. Essentially, this was the design principle followed in the Enfield Autocar three wheeled "Autorette" motor car of 1912.

LEFT-BOTTOM: Diagram from patent number 1633 taken out in 1912 by William Allday and Charles Edward Simms for a pneumatically operated starter for motor car engines. Working on the rack and pinion principle, compressed air stored in a suitable container on the vehicle was used to propel the rack, which was attached to an air operated piston, so turning the pinion attached directly to the engine crankshaft. On the return stroke of the piston/rack the free-wheel device built into the pinion ensured that no engagement between the starter and engine crankshaft took place.

ILLUSTRATION

RIGHT: This highly elaborate illustration shows the company letterhead used in connection with correspondence from the Matchless Works in Fallows Road in 1910. Interestingly, the Fallows Road site is described as being of 3 acres and the Sydenham Road site being of 4 acres. These were rather generous calculations on someone's part as the Fallows Road site was approximately 1.96 acres and the Sydenham Road site no more than 2.75 acres at this time.

The two signatures at the top of this page are of William Allday in 1910 and Percy William Allday, the son of Edmund Allday, in 1918.

ALLDAYS

ALLDAYS & ONIONS LTD
BIRMINGHAM & LONDON
ALLDAYS & ONIONS LTD
BIRMINGHAM & LONDON

PHOTOGRAPHS

LEFT: Two views of the 1910 Alldays 14/18 h.p. four cylinder engine which had a bore of 3.375ins diameter by 4.25ins stroke. The engine featured force feed lubrication by a rotary pump, carburation by a float feed spray type carburettor and high tension magneto shaft driven from the timing gears.

From an early photograph it would appear that production engines were run-in by being belt driven from an overhead layshaft before being started up and run on petrol. However, the test house appeared to be devoid of any kind of dynamometer to put the engines under load, or to enable checks to be made of their actual output in terms of horse power developed.

BELOW: The 1910 Enfield 18/22 h.p. four cylinder engine with a bore of 95mm diameter and stroke of 115mm. Like the Alldays engines these featured cylinders cast in pairs which were hydraulically tested for porosity after machining. Pistons were of close grained cast iron and the exhaust side of this engine featured four individual down pipes direct from the cylinder block into a cast iron horizontal common manifold which then connected directly to the exhaust system. Although the Enfield and Alldays engines were being built in the same factory they retained their individual design features, as can be seen from these photographs.

engine, but based on a 8ft 6ins wheelbase chassis; a 18/22 h.p. model which used the Enfield 4 cylinder engine with a bore of 95mm and stroke of 115mm and based on a 9ft 3ins or 9ft 9ins wheelbase chassis and a 30/35 h.p. model which used a massive 4 cylinder engine with a bore of 120mm diameter and stroke of 135mm giving it a capacity of 6.1 litres. This car had a 3 speed gearbox and was based on a 10ft wheelbase chassis. Prices for the Enfield cars ranged from £201-12s-0d to £567. The Alldays range of cars for 1911 now listed only one 10/12 h.p. 2 cylinder model, fitted with a 4 speed gearbox, 4 seater body and based on a 8ft 6ins wheelbase chassis priced at £262. No less than three new engines had been introduced, a 12/14 h.p. 4 cylinder with a bore of 76mm diameter and stroke of 120mm, a 16/20 h.p. 4 cylinder with a bore of 86mm diameter and stroke of 130mm and a 25/30 h.p. 4 cylinder with a bore of 100mm diameter and stroke of 130mm. These were used to power two 4 seater cars and a 6 seater car, based on chassis with a wheelbase of 9ft 0in for the 12/14 h.p. and 9ft 6ins for the two larger models with prices at £294, £367 and £409 respectively. The 7 seater 30/6 cylinder 35 h.p. model was now based on a 10ft 4ins wheelbase chassis and cost £567. The company now had 8 different motor car engines in production, all the casting, machining and assembly being undertaken in its own factories. All vehicle assembly work was undertaken in the Matchless Works and now included vans, lorries, light carriers, omnibuses, charabancs, fire tenders and motor rail cars.

A special 20 h.p. railcar to carry 20 passengers had been supplied to the Egyptian State Railway in 1911 which had a turntable built beneath the chassis frame fitted with a jacking mechanism which allowed the car to be jacked up clear of the rails at its destination,

M.R.
M.R.
ALLDAYS & ONIONS
BIRMINGHAM & LONDON.

ALLDAYS & ONIONS LTD
BIRMINGHAM AND LONDON.

PHOTOGRAPHS

LEFT: The two photographs to the left show similar railcars, one fitted with a canopy and the lower one completely open, but note the waterproof sheets provided, draped across the seats to provide some degree of protection from the elements. The lower picture shows a railcar similar to one supplied in 1905 to the London, Brighton & South Coast Railway as an engineer's inspection car.

BELOW: The photograph shows a 6 to 9 passenger railcar built in 1912 which could be driven from either end and, like trams, had seat backs which could be moved backwards or forwards to enable the passengers to sit facing the direction of travel. Railcars were available with 3.5 h.p., 10 h.p. or 14 h.p. engines. A 3.5 h.p. model with 2 speed gearbox, but no reverse, cost £95 to £105 and 10 h.p. models with 2 speed gearboxes and forward and reverse gears cost £235 to £285.

Railcars designed to carry passengers and goods were also available. A 14 h.p. railcar to carry passengers and a 1 ton load cost £355. Other passenger/goods models able to carry loads of 2 tons and 3 tons were fitted with three speed gearboxes, plus forward and reverse gears. (Photograph courtesy of Birmingham Library Services)

turned through 180°, lowered back onto the rails, and then driven back to its starting point. Later, in October 1911, a 10 h.p. 4 seat railcar with a wheelbase of 4ft 10ins and track of 2ft 6ins with engine number 4503 and chassis number 4086 was built for a company called Strain & Robertson. In October 1912 two 10/12 h.p. railcars having a wheelbase of 6ft 9ins and track of 5ft 6ins were despatched to the Paapin Railway Company in Buenos Aires, Argentina. These had a combined gearbox and axle assembly, were fitted with four revolving seats covered by a canopy and were chassis numbers 4402 and 4404 with engine numbers 4931 and 4393 respectively. One month later another railcar was despatched to South America, this time to Francisa de Lima, Peru. This model was fitted with a 14/18 h.p. engine, number 5093, chassis number 4680, had a 6ft 3ins wheelbase, track of 5ft 5.75ins and was fitted with a GPO type gearbox and Hele-Shaw clutch. In the early months of 1912 two more 10/12 h.p. railcars with a 6ft 9ins wheelbase and track of 5ft 6ins and fitted with four revolving seats were despatched to the company Grossmith & Luboldt. These carried chassis numbers 4406 and 4408 and engine numbers 5243 and 5245 respectively. One railcar, at least, has survived into preservation, County Donegal Railways Number 1 was supplied in 1906 as an inspection car. This open topped 6 seater 10 h.p. model had originally cost £237. In 1920 the original engine was replaced by a Ford petrol engine and the car was covered in. This railcar is now in the Belfast Transport Museum.

The financial results for the year 1910/11 were disclosed to shareholders in a meeting held in "Great Western Hall" on the 30th November 1911. A net profit of £15,357-17s-0d was announced and, interestingly, the balance sheet shows that the company still owned

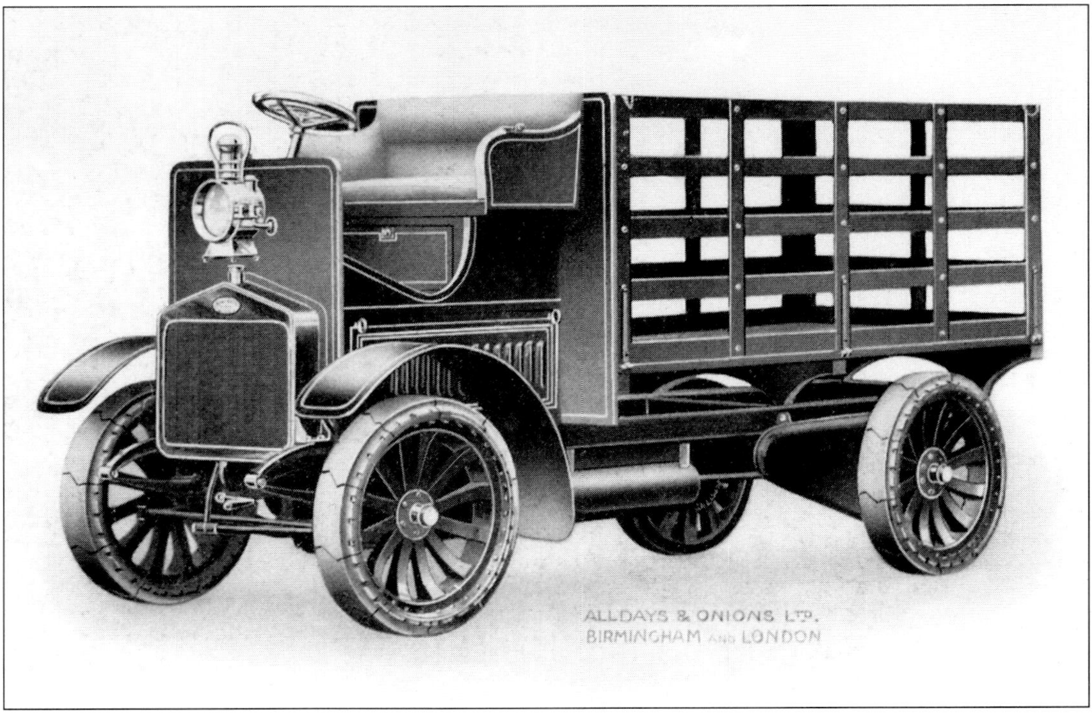
ALLDAYS & ONIONS LTD.
BIRMINGHAM AND LONDON

the premises in Branston Street and Commercial Street which were valued at £4,962-14s-1d. The Great Western Works was valued at £33,966-10s-8d and the Matchless Works at £24,171-11s-1d. Plant and machinery at the two works was valued at £59,299-14s-0d.

In 1912 Britain was in the grip of a crippling coal miners' strike which was rapidly bringing some sectors of industry almost to a halt. Edmund Allday, however, was touring India on a sales drive to get orders for the company, but was kept well informed on matters at home through regular correspondence from brother William and son Percy. The miners' strike was already affecting the Alldays factories when William wrote to Edmund wishing that he 'could get better delivery of cars and motor cycles', but the factory had just started going in later at 8.00a.m. each day and the works were closed every Saturday. Percy wrote in more detail about the effects of the strike which was making it difficult to keep the boilers and three steam engines at the works under full power with low grade coal having to be used. However, he had just ordered two boats (canal barges which could unload at the company's own wharf) of cobbles which had cost 16s-6d and £1-12-6d respectively in the hope of keeping the works going for a little longer.

The commercial vehicle range for 1912 was extended by the addition of a new 3 tonner which, initially, was powered by a 35 h.p. 6 cylinder engine with a bore of 95mm diameter and stroke of 115mm fitted with a White & Poppe carburettor. This model had a four speed gearbox with chain drive to the rear axle via a transverse cross-shaft. The first 3 tonner to be built was exported to Argentina. By the end of 1912 the commercial vehicle range covered 5 cwts to 5 tons, plus nine different models of railcars, including types to carry passengers and goods in 1 ton, 2 tons and 3 tons capacities powered by engines of 3½ h.p., 10 h.p., 14 h.p., 20 h.p. and 30 h.p. and with prices ranging from £95 to £355. The specifications for the commercial vehicle range were as follows:- 5 cwt 3 wheeler powered by a single cylinder engine with a bore of 100mm diameter and stroke of 115mm and fitted with a 2 speed gearbox. This became known as the "Expressodel" Autocarrier and was built following the basic patented design of Charles Edward Simms. The engine was usually to the rear driving the single rear wheel via a chain. The driver sat high above the engine with the load carrying box, which measured 3ft 6ins by 3ft 6ins by 2ft 6ins, above the front steering axle and immediately in front of the driver. An alternative design to this was also offered where the engine was mounted almost centrally in the chassis and the driver sat above it steering the front wheels which were just in front of his feet and the load box was to the rear and partly overhanging the single driving rear wheel. The price of this unusual model was £100, finished in Alldays standard green colour, plus £3 for a casing to enclose the chain drive and another £2-10s-0d for a pair of side and tail lights. A 5 cwt four wheeler based on "The Midget" cyclecar chassis was powered by a 2 cylinder 8/10 h.p. engine with a bore of 86mm diameter and stroke of 92mm fitted with a 3 speed gearbox and priced at £130. 10 cwts 4 wheeler was powered by a 10/12 h.p. 2 cylinder engine with a bore of 95mm diameter and stroke of 115mm, fitted with a 3 speed gearbox and priced at £240. 10 cwts model was powered by a 12/14 h.p. 4 cylinder engine with a bore of 76mm diameter and stroke of 120mm, fitted with a 4 speed gearbox and priced at £285. The 15 cwts model was powered by the same 4 cylinder engine as the previous 10 cwts model and priced at £295.

PHOTOGRAPHS

TOP: The 1910 Enfield 30/35 h.p. Model 301 Limousine which cost £567 with body upholstered in a choice of French cord or Morocco leather. The four cylinder engine had a bore of 120mm diameter and stroke of 135mm giving it a capacity of 6.1 litres, and drove through a leather cone clutch to a 3 speed gearbox and then to a gear driven rear axle. This motor car was based on a 10ft wheelbase chassis frame with an overall length of about 14ft and overall width of 5ft 6ins. A Double Landaulette body, Model 302, was also available on this chassis at £546. In chassis form the 30/35 h.p. model could be had for £367-10s-0d, plus tyres.

BELOW: A 14/18 h.p. Alldays forward control lorry rated at 20/30 cwts built in 1912 for the Uganda Company. This was supplied with three single axle trailers with similar stake sided bodywork to that shown on the lorry and used to transport cotton and wool. This illustrates just how versatile the company was, not only being able to offer a huge range of standard products, but also to be in a position to build 'one-offs' to suit an individual customer's requirements.

Royal Mail.
ALLDAYS & ONIONS Ltd

Royal Mail
ALLDAYS & ONIONS Ltd
BIRMINGHAM AND LONDON

PHOTOGRAPHS

LEFT: Two examples of light delivery vans supplied to the G.P.O. for the Royal Mail service. A 10/12 h.p. 10 cwts model and a 14/18 h.p. 15 cwts model, both chain driven for heavy duty service. In 1910 a 20 h.p. van, chassis number 2152CD, with engine number 2672 and fitted with Clincher solid tyres had also been supplied to the G.P.O. and in 1913 three 25/30 h.p. vans, chassis numbers 4632, 4634 and 4636 had also been supplied, the first van on a 10ft 6ins wheelbase chassis and the latter two on 10ft 3ins wheelbase chassis. All three were fitted with chain drive, 3 speed gearboxes and pneumatic tyres.

The general range of commercial vehicles included Model 16 10/12 h.p. 10 cwts at £241-10s, Model 17 14/18 h.p. for 15 cwts at £357, Model 45 20/24 h.p. for 20 cwts at £472-10s and Model 46 20/24 h.p. for 30 cwts with twin solid tyres on the rear at £524.

BELOW: A 30 cwts Stores lorry supplied to the G.P.O. in 1910. This vehicle was purchased for trials, together with a 16/20 h.p. Ryknield and a 20 h.p. Halley. The Alldays machine was found to be the best of the three and five more were ordered by the G.P.O. (Photograph reproduced courtesy of British Telecommunications Plc.)

The 1 tonner was powered by a 16/20 h.p. 4 cylinder engine with a bore of 86mm diameter and stroke of 130mm, fitted with a 4 speed gearbox and priced at £340. Both the 30 cwts and 40 cwts models were powered by 25/30 h.p. 4 cylinder engines with a bore of 100mm diameter and stroke of 130mm, fitted with 3 speed gearboxes and costing £430 and £460 respectively. The 3 tonner and new 5 ton models were powered by 30/40 h.p. 4 cylinder engines with a bore of 120mm diameter and stroke of 140mm and were fitted with 4 speed gearboxes and cost £590 and £680 respectively.

As well as motor vehicles the company introduced specially designed adjustable engine stands for use in garages and motor manufacturing workshops which allowed an engine to be bolted to the stand and rotated allowing work to be carried out to any part of the engine and they also introduced small portable cranes for use in workshops.

One interesting little car to be introduced into the Enfield range in 1912 was a three wheeler known as the "Autorette" with two front steerable wheels and a single rear wheel driven by a single cylinder engine. This was the motor car patented in 1911 by Charles Edward Simms described simply as a mechanical engineer and designer working at the Matchless Works and which formed the basis for the three wheeled commercial "Expressodel".

With Alldays being well established in the field of pneumatic engineering it was of little surprise to find the company introducing a compressed air self-starter for motor cars working on a rack and pinion principle and patented by William Allday and Charles Edward Simms in 1912.

ALLDAYS "MIDGET" CAR

ALLDAYS & ONIONS LTD
BIRMINGHAM & LONDON.

Some company records exist detailing motor vehicle production for the period 1907 to 1913, but all the chassis numbers quoted are even numbers, suggesting that there were probably other ledgers containing the uneven chassis numbers. If this is so, and it is extremely unlikely that vehicles were built with only even chassis numbers, then the production figures quoted should be doubled. In 1907 86 vehicles were built rising to 109 in 1908. The records for 1909 and 1910 production are obviously incomplete and therefore are not worth quoting, but 127 vehicles were built in 1911 rising to 278 in 1912 with, again, incomplete figures for 1913. Doubling the figure for the most prolific year, 1912, indicates a production of about 10 vehicles per week and this would be quite a realistic figure considering the production facilities available and the types of vehicles under construction. One also has to remember that motor vehicle production was just a part of the company's output, having to share facilities being used for the production of cycles and the enormous range of engineering products. The most popular size of vehicle in 1912 appears to have been the 12 h.p. model with 132 vehicles, followed by the 16 h.p. model with 65 vehicles and the 14 h.p. model with 36 built. The most popular motor car body style was the Torpedo with 126 built, followed by 30 Victoria's and 26 chassis supplied to dealers for bodies to be added later. These figures include all types of vehicles, motor cars, vans, lorries, charabancs, buses, trolley buses and rail cars.

The motor car range for 1913 was similar to that of 1911, but now included 11 models, some of which were modified forms of the 1911 cars. The 16/20 h.p. model was made available on two different chassis with a wheelbase of 9'-0" and 9'-9" which cost £290 and £300 in chassis form and the 25/30 h.p. model was available on a 10'-6" wheelbase chassis as well as the 9'-6" wheelbase chassis, the longer wheelbase model costing £350 in chassis form. The top of the range Alldays 30/35 h.p. 6 cylinder model was now being offered on a longer 11'-6" wheelbase chassis and cost £450 as a chassis or £570 for a complete 7 seater car.

For 1914 the motor car range was slimmed down to basically five models, a 8 h.p. 2 cylinder 3 seater based on a 7ft 0in wheelbase chassis priced at £130, a 8/10 h.p. 4 cylinder 3 seater on a 8ft 0in wheelbase chassis priced at £165, a 12/14 h.p. 4 cylinder 4 seater on a 9ft 3ins wheelbase chassis at £350, a 16/20 h.p. 4 cylinder 6 seater on a 10ft 0in wheelbase chassis priced at £400 and a 25/30 h.p. 4 cylinder 6 seater car based on a 10ft 6ins wheelbase chassis priced at £485. All these motor cars featured a three speed gearbox. The first two models in the range were more popularly known as the "Midget" and featured a chassis frame made in a 'U' shape from a single piece of channel section rolled steel. Both models also featured cantilevered semi-elliptic road springs, leading at the front and trailing to the rear, a design which was to be tried again in the ill fated "Bullet" introduced after the First World War. The 5 cwts light delivery van based on the "Midget" chassis was called the "Expressodel" Number 3 and, to add to the confusion, a tradesmans motor cycle and sidecar outfit was also being offered called the "Expressodel" No.2!!

Motor cars could be ordered from the works by sending "a substantial deposit" with the order. Delivery could be made by an "expert driver" at a cost of 12s-6d per 10 hours day, plus board and lodgings/travelling expenses as and when required!!

PHOTOGRAPHS

TOP: The forward control stake sided lorry built in 1912 for the Uganda Company photographed in Fallows Road complete with three stake sided single axle trailers.

CENTRE: The "Midget" motor car with its cantilevered road springs clearly visible in this picture, which also shows it to be quite a pretty little car.

BOTTOM: The 5cwts van based upon the "Midget" motor car chassis and known as the "Expressodel" number 3 light delivery van.

LEFT-TOP: It was this William Allday, born in 1851, who was responsible for the union of the Alldays and Onions companies in 1885. William was the real driving force behind the new company and he also initiated most of the patents taken out by the company. A patent taken out by William in 1886, for improvements to portable forges, describes him as William Allday the younger and his address was given as 335 Coventry Road, Small Heath, just a few minutes walk from the Sydenham Road works.

LEFT-BOTTOM: William's younger brother, Edmund, was born in 1855 and died in 1935. Edmund was also involved in taking out patents, but from family correspondence it would appear that he was much happier travelling the world getting orders for the company, than actually sitting behind a desk overseeing the day to day running of the works. There are details in the Appendix of this book of an interesting patent involving Edmund who co-designed a very early form of breathing apparatus for removing poisonous gases from the lungs of miners and sub-mariners!

RIGHT-BOTTOM: Percy William Allday, born in 1882 and the son of Edmund, resigned in 1921 after assisting his father to run the company following the death of his uncle in 1916. Percy went on to establish his own company, William Allday & Co., with offices in Paradise Street, Birmingham, making a similar range of foundry equipment to that of Alldays & Onions. Percy Allday died in 1937.

BELOW: A 30 cwts Alldays lorry supplied to the Naval Store Dept, with bodywork believed to be by Edward Evans (Coachbuilders) Ltd. of Dudley.

Alldays commercial vehicles were to be found all over Britain and Ireland being operated by such people as Kynoch Limited in Birmingham who operated 2 ton and 5 ton trucks, Claude Odlum, a flour miller of Portarlington, Ireland, who operated a 5 tonner and the Derby Co-operative Society who had eleven Alldays vehicles in their fleet.

A 25/30 h.p. 20 passenger motor bus was operated by Ernest W. List between Debenham and Ipswich and the bodywork was designed in such a way so that it could be used for furniture deliveries between trips carrying passengers. Other motor buses based on the 30/35 h.p. 3 ton chassis were in service with the Mid-Cheshire Motor Bus Co.Ltd. Brewing firms like Alton & Co. Ltd. of Derby and Hancocks Brewers also used Alldays trucks and vans, the latter company taking delivery of six 3 ton vans in June 1915. In Birmingham the Corporation took delivery of a 30/40 h.p. 5 tonner fitted with a 1,000 gallon water tank and spraying equipment which could deliver a spray up to 50 feet wide and used for street cleaning.

War work meant that motor car production was concentrated at the Waverley Works with production of agricultural tractors, motor cycles, cycles and commercial vehicles, which were in great demand, concentrated at the Matchless Works, whilst general engineering work and the production of shell nosings, etc., was concentrated at Great Western Works.

On Saturday September 2nd 1916 William Allday, aged 66 years, died following an operation. Edmund Allday, his younger brother, then took over the reins of the company as Managing Director and was to be ably assisted by his son Percy William Allday.

Even during the war the company was still offering the "Light Four" range of vehicles which comprised an open 2 seater at £215, a 2 seater coupé at £255, described as "The car for the Professional Man", and a 5cwt delivery van at £170. All these models were based

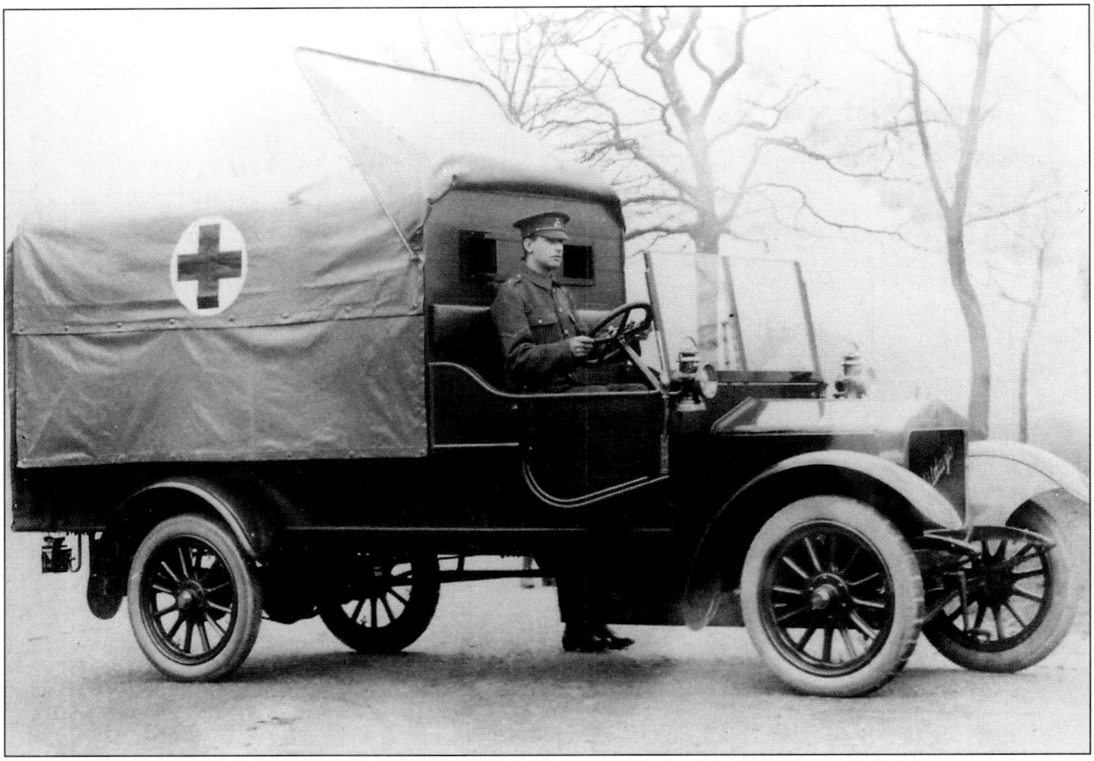

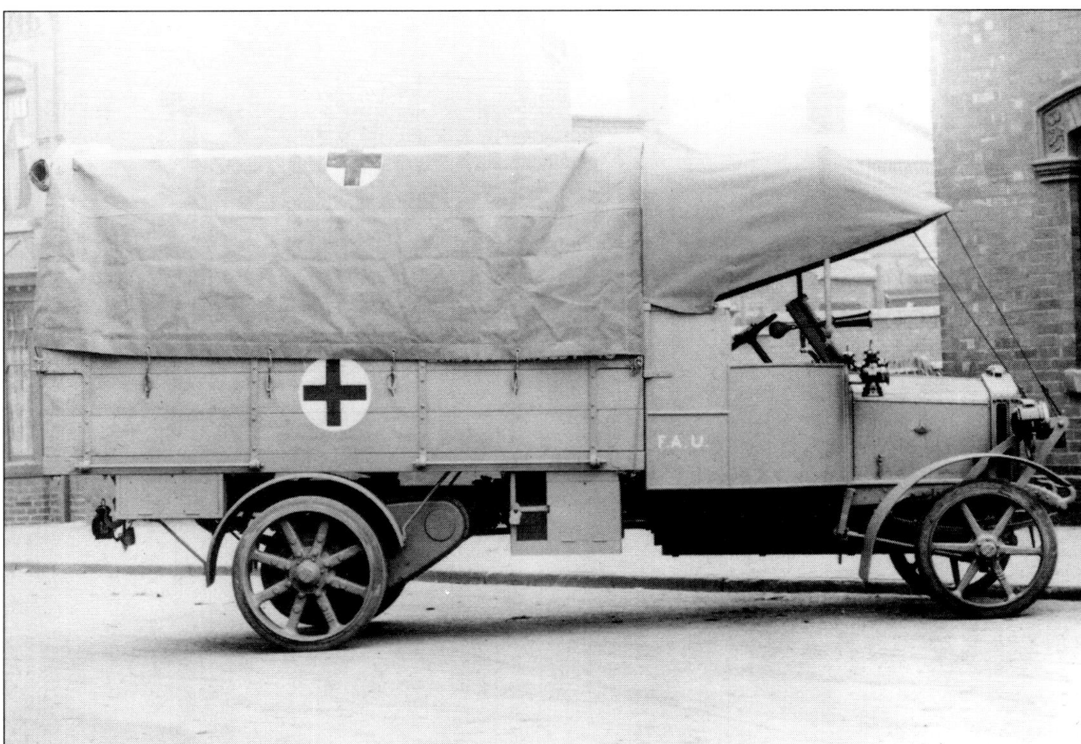
F.A.U.

PHOTOGRAPHS

LEFT: These photographs show two military ambulances, built during the First World War, based upon two very different chassis. The upper picture is of an ambulance based on a long wheel-base motor car chassis whilst the vehicle in the lower picture was based on a 30 cwts - 40 cwts commercial vehicle chassis with chain drive.

(Photographs courtesy of Birmingham Library Services)

BELOW: Based around the Alldays patent number 25189, this "Expressodel" Number 1 Autocarrier was capable of carrying 5 cwts and cost £100 in basic form. The engine, beneath the driver's seat, was a single cylinder 7/8 h.p. water cooled type with a bore of 4" diameter and stroke of 4½". Cooling was via duplex radiators, one of which is clearly visible in this picture, and the chain drive to the rear wheel was via a gearbox giving 2 forward and 1 reverse speed. Braking was via foot pedal to internally expanding brakes on the front wheels and by a hand lever to the rear driven wheel. The fully enclosed chain drive shown here cost an extra £3 and the plated rail on top of the load carrying box cost an extra £1-15s-0d. The load carrying box measured 3ft 6ins by 3ft 6ins by 2ft 6ins, made to 'Best Coach Finish' and painted in Alldays standard green. Strangely, an alternative "Expressodel" was also available with the engine moved to an almost central position and the box above the single rear driving wheel and with the steering column moved to above the two front wheels.

upon a single piece chassis frame with a wheelbase of 8ft 1½ins and front track of 4ft. The four cylinder engine had a bore of 61mm diameter and stroke of 100mm driving through a leather cone clutch to the 3 speed gearbox and worm drive rear axle.

In June 1918 Mr N.A.Hilton, a member of the Alldays staff, was elected to the Council of the Society of Motor Manufacturers and Traders for the year 1918/19. In January 1919 the company shareholders looked forward to a 5% dividend on their Ordinary shares and 6% dividend on their 'A' shares.

At the end of the war the Enfield Autocar Company name was discontinued and all motor cars were then produced under the name of Enfield-Alldays Motors Ltd., which, together with "Allon" motorcycles, were concentrated at Waverley Works.

Development work had commenced on the revolutionary 10 h.p. air cooled radial engined motor car, but the company were still offering a 15 h.p. model, of conventional design, powered by a new 6 cylinder 2.5 litre engine with a bore of 70mm diameter and stroke of 110mm which developed 21 h.p. at 1,250 r.p.m. and 52 h.p. at 3,000 r.p.m. Fitted with a four speed gearbox on a 9'-9" wheelbase chassis the car was priced at £475 complete with touring car body. However, although advertised in the motoring press it is thought that this car was never actually put into production.

Just prior to the war plans had been instigated to put the 12/14 h.p. car into volume production and all the necessary jigs and fixtures had been made. However, this decision was overturned after the war, to concentrate production on "The Bullet", and all this expensive equipment was sold off to William Morris ostensibly as 'scrap metal'!! Apparently Alldays also supplied rear axles to Dennis Brothers Ltd., so they also contributed to other companies in the British motor industry.

ANTIPODEAN ALLDAYS

Alldays foundry and general engineering products were exported all over the world, but it was surprising to find their motor cars reaching the other side of the world within a very short time of going into production. However, early incursions were made into Australia, Tasmania and New Zealand, through agents such as Dalgety & Co.Ltd., of Sydney, Behrens & Marshall of South Australia, A.G.Webster & Sons of Hobart, Tasmania, Messrs Howie & Company of Auckland, A.F.Cole of Hawera, Hawkes of Christchurch and The Wellington-Wairarapa Motor & Cycle Co.Ltd., of Masterton, New Zealand.

In Australia there was the report of a an Alldays "Traveller" being sold in 1902 by Tarrant Motors in Melbourne and a 7 h.p. single cylinder model, owned by a Mr. G.T.Holmes, took part in the 1905 Dunlop Reliability Motor Contest which was run between Melbourne and Sydney. In 1906 an Alldays car was an entrant in the Australian National Park Hill Climb, but trouble with the sparking plugs caused it to drop out of the actual contest, but afterwards the car was allowed to tackle the challenging hill climb, which it completed splendidly with no less than seven people on board! In the same year a Dr. A.P.E. O'Leary of Glenelg, South Australia, was using a 10/12 h.p. Alldays and in Tasmania a 2 cylinder 10 h.p. model was in the ownership of a Mr. P.O. Fysh of Launceston.

Early examples of these new Birmingham made motor cars also appeared in New Zealand such as one owned in 1907 by a Mr J.Scott of Marlborough who later converted it into a motor van with a locally built body. One of New Zealand's pioneering motorists, Duncan Rutherford, actually owned two Alldays cars, one of which, a 10 h.p. model, became the very first motor car to cross a bridge, built by Rutherford himself, over the Hanmer River in 1908. A much later car, a 1913 2 cylinder "Midget" is believed to have been owned by a farmer near Wanganui and another 1914 four cylinder "Midget" was owned by a Dr. Milroy of Hawera. These were obviously just a few of the cars exported to New Zealand.

Company records show two 10 h.p. models exported to Dalgety of Australia in 1907, both fitted with factory built bodies, and ten despatched to the same agent in 1908, all being 10 h.p. models, one with a 4 seater Phaeton body in green and one being a 2 seater Victoria finished in blue and three in chassis form A chain driven 14 h.p. "Colonial" model are also shown as having been delivered to Dalgety's in 1909. Hawkes, of New Zealand, are shown to have taken 11 cars in 1908, one 8 h.p., one 20 h.p. and the remainder being 10 h.p. models. All were fitted with factory built bodies which included a Roi de Belge finished in green, two 4 seaters with a swinging front seat and a 2 seater. The motor cars supplied to Hawkes were in a variety of different colours which included green, dark green, red, cream, and white. The Wellington-Wairarapa Motor & Cycle Co. Ltd. are shown to have taken two 10 h.p. models in 1908, both with factory built bodies and both fitted with chain drive. Later models to reach Australia included a 1911 four cylinder 14/18 h.p. model owned by Mr.T.A. Brock of Woodville, who was a shipping agent in Port Adelaide. Enfield models also found their way to Australia, such as a 12 h.p. 4 cylinder model owned by Mr H.J. Adamson of Forestville, Adelaide, in 1913 and a 1914 "Autolette" 2 cylinder 8/10 h.p. model owned by a Mr. R.G. Cheeseman of Kilkenny.

PHOTOGRAPHS

TOP: This 8-10 h.p. Alldays motor car looks rather worse for wear after being used for some pioneering motoring, about 1905, in New Zealand. Owned by Ivan Wintringham and Bert Mitchell, seen here with the car, it was used to tackle a journey from Blenheim to Tophouse before any roads had been constructed. The pioneering pair had to hack their way through bush, drove through rivers, pushed and dragged the car, and sometimes had to use a block and tackle to prevent it sliding down into ravines!

(Photograph courtesy of Rob Knight - New Zealand)

BOTTOM: In New South Wales, Australia, this little Alldays railcar was used on the 24" gauge railway built during the construction of the Burrinjuck Dam. The railcar is thought to have entered service on the 25 mile line in 1908 and ran on this line during construction work up to about 1929 when the line was taken up and all the rolling stock sold off. Like most other Alldays railcars the seat backs could be repositioned to enable the vehicle to be driven in both directions when, as on this particular line, there was no means of turning the railcar around at the end of its journey.

(Photograph courtesy of John Newland - Australia)

ERNEST W. LIST,
DEBENHAM.
BJ 2052

MUSTERS RD & MELTON

CHARABANCS & BUSES

During the period 1911 to 1913 the company records show that the production of specialised charabancs and buses was far from being what might be termed 'one-offs' and these models were built in some numbers and chiefly for one particular dealer, McIlwraith, McEachern & Co., of London.

The basic mechanical configuration was a chain driven chassis, using Birmingham made Brampton chains, and a three speed gearbox with whatever wheelbase chassis frame the customer required and, initially, fitted with solid tyres, single at the front and double at the rear.

The suspension was, of necessity, heavier duty than normal and early 20/24 passenger models were fitted with front springs with 8 leaves 1.75" wide and rear springs with 12 leaves 2" wide. This rapidly changed to slightly stronger rear springs with 12 leaves 2.25" wide. On the smaller 16 seater models the standard suspension used was 8 leaves 1.5" wide on the front and 9 leaves 1.75" wide on the rear.

In 1911 one bus and six charabancs were built. The bus, chassis number 4144AE3, was a 24 passenger model powered by a 30/35 h.p. engine, number 4537, on a 13ft wheelbase chassis with 6 rows of seats. The first four charabancs were 20 h.p. models, a 24 seater with six rows of seats and three 20 seaters with five rows of seats, all based on 12ft wheelbase chassis and all featuring enlarged radiators to assist engine cooling under arduous conditions. The chassis numbers were 4056ADV2, 4060ADV2, 4062ADV2 and 4064ADV2 and engine numbers, 4491, 4555, 4551 and 4553 respectively. The last two charabancs were a 20/25 h.p. 20 passenger model with five rows of seats on a 10ft wheelbase chassis and fitted with a special GPO 3 speed gearbox, being chassis number 4156ADV3 and engine number 4487, and a 16 h.p. model with a 16 seater body on a 10ft wheelbase chassis, number 4246, and engine number 4661. The latter model was delivered to Martin & Co.

In 1912 twelve charabancs, one chassis, and five buses were built starting with four 20/25 h.p. 24 passenger charabancs in January/February, based on 12ft wheelbase chassis frames, with GPO type 3 speed gearboxes specified. These were chassis numbers 4158 ADV3, 4160ADV3, 4162ADV3 and 4164ADV3. These were followed in March by two similar 24 seater charabancs with six rows of raised seats, chassis numbers 4290EV1 and 4294EV1. In May two more 25/30 h.p. 24 seater charabancs were built on chassis with 12ft 3ins wheelbase, chassis numbers 4296EV1 and 4450EV2. In June/July two more 25/30 h.p. charabancs were built, one on a 12ft wheelbase chassis, number 4454EV1, and one on a 12ft 3ins wheelbase chassis frame, number 4458EV2.

One chara-banc chassis and one complete chara-banc were built in August, the chassis being a 30/35 h.p. model on a 13ft 2ins wheelbase suitable for a 30 seater body, chassis number 4150AE3, supplied to Tozer, Kemsley & Fisher Ltd. The complete vehicle, chassis number 4524EV2, was a 25/30 h.p. model on a 12ft 3ins wheelbase chassis and was fitted with six rows of wooden lath seats 68" wide. The final chara-banc to be built in 1912 was also a 25/30 h.p. model on a 12ft 3ins wheelbase chassis, but fitted with pneumatic tyres and was chassis number 4526EV2.

PHOTOGRAPHS

TOP: This 25/30 h.p. 20 seat vehicle was put into service in 1914 by Ernest W. List of Debenham, Suffolk, and used for a local bus service to Ipswich. The body was constructed in such a way as to enable the vehicle to be used as a delivery van, or furniture van, when not being used for passenger work and would, today, be more popularly known as a 'Combi'.

(Photograph courtesy of Birmingham Library Services)

BOTTOM: Based on a 35/40 h.p. 3 ton chassis this double deck bus was put into service by the West Bridgford Urban District Council in 1915, and was just one of the seven Alldays buses to be operated in their fleet. The bodywork is thought to be the work of a specialist body builder, Starey & Wooley, and not a factory built body.

(Photograph courtesy of Birmingham Library Services)

ALLDAYS

MOTOR CYCLES.

Alldays Patent
'ALLON'
Two - stroke.

Fast on Hills.
Light on Bills.
The Best
Lightweight
Manufactured.

TOURING CARS.

10 H.P.
LIGHT CARS.
12 14 H.P. 16 20 H.P.
and 25 30 H.P.
Four or Five-seated
Touring Models.
The most up-to-date and fully equipped cars on the market.

COMMERCIAL VEHICLES.

LORRIES, VANS,
CHAR-A-BANCS,
OMNIBUSES,
TIP WAGONS,
&c.
For loads of 5 cwt.
to 5 tons.

Estimates for any type of Commercial Cars on request. Write for catalogues.

ALLDAYS & ONIONS Pneumatic Engineering Co., Ltd.,

Matchless Works, BIRMINGHAM, and 58, HOLBORN VIADUCT, E.C.

Cable Address—ALLDAYS, BIRMINGHAM. 36063

Established 1650.

The buses built in 1912 started in April with a 16/20 h.p. model on a 10ft wheelbase chassis fitted with a 16 passenger body with four rows of seats each 56" wide and was chassis number 4356 CV1. In May a 25/30 h.p. model on a 12ft 3ins wheelbase chassis was built with 6 rows of seats, chassis number 4298 EV1. This was followed in June by a similar vehicle, but based on a 12ft wheelbase chassis and with six rows of seats 70" wide, being chassis number 4452EV2. In October/November two more 25/30 h.p. buses were built, a 12ft 3ins wheelbase model with six rows of seats, chassis number 4528EV2, and a 12ft wheelbase bus with five rows of seats on chassis number 4620 EV3.

The production records for 1913 are incomplete, but details are shown of four buses built during the period January to March, three of these being 25/30 h.p. models on 12ft 3ins wheelbase chassis and all fitted with bodies having five rows of seats 68" wide. Chassis number 4622EV3 was fitted with solid tyres, but chassis numbers 4628EV3 and 4630EV3 were fitted with pneumatic tyres. A 16/20 h.p. bus based on a 9ft wheelbase chassis was built in January fitted with a direct drive 4 speed gearbox, pneumatic tyres and a body with three rows of seats 62" wide, this being chassis number 4716CV3.

At least one double decker bus is known to have been put into service by 1915, but it is not known whether this was bodied at the works, or by a specialist bodybuilder.

ILLUSTRATION

LEFT: A 1916 advertisement covering all types of road vehicles currently in production.

PHOTOGRAPH

BELOW: Based on a 3 ton 30/35 h.p. long wheelbase Enfield chassis this single deck motor bus was put into service in April 1916 by the Mid-Cheshire Motor Bus Co.Ltd.

(Photograph courtesy of Birmingham Library Services)

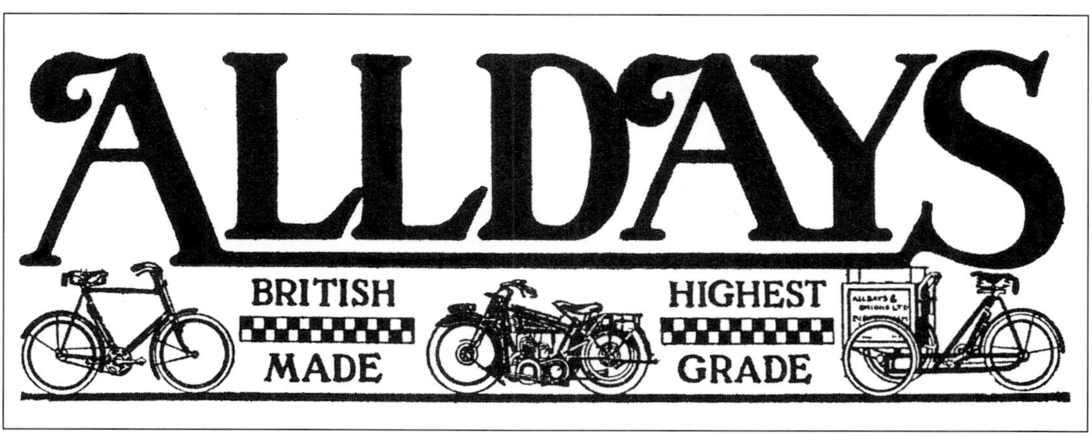

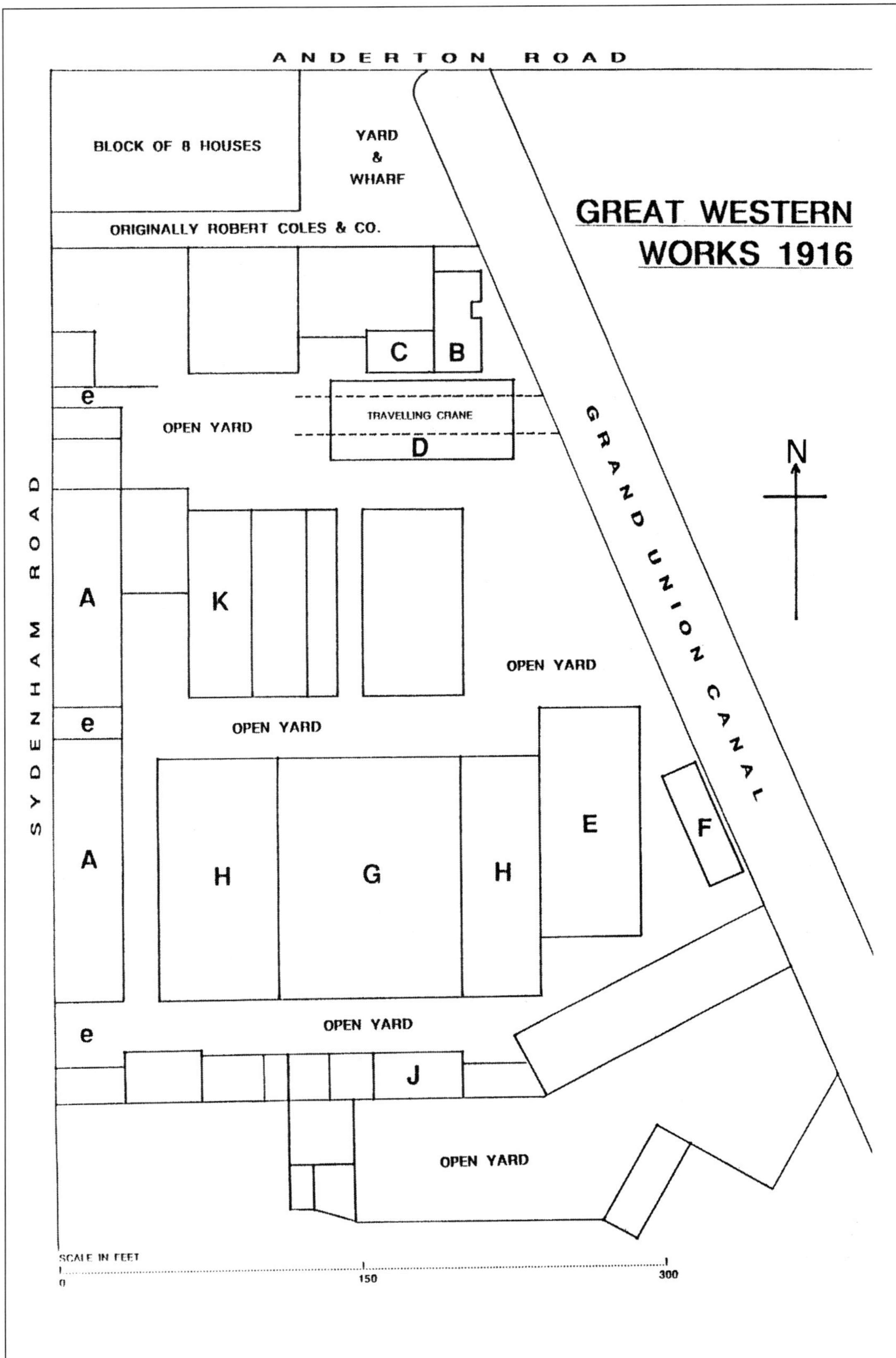
GREAT WESTERN WORKS 1916
ANDERTON ROAD
SYDENHAM ROAD
GRAND UNION CANAL
BLOCK OF 8 HOUSES
YARD & WHARF
ORIGINALLY ROBERT COLES & CO.
C
B
e
OPEN YARD
TRAVELLING CRANE
D
N
A
K
OPEN YARD
e
OPEN YARD
E
F
A
H
G
H
e
OPEN YARD
J
OPEN YARD
SCALE IN FEET
0
150
300

EXPANSION OF THE GREAT WESTERN WORKS AND ITS PRODUCTS

Just as the company's catalogue of products expanded so did the factories to produce them. In 1900 part of the travelling crane which ran alongside the Foundry and Core Shop in the Great Western Works was enclosed to provide a covered work area and storage area when an open shed was erected measuring 90 feet long by 33 feet 6 inches wide. The crane was mounted on rails 20 feet apart and was used to unload raw materials coming to the works by barges using the Grand Union Canal and may also have been used to load barges with goods produced at the works for distribution around the country.

Lifting tackle, hoists and cranes were being sold under the name of Eades Hoist Co.Ltd., but everything was being made at Sydenham Road. This range of specialised products included simple chain pulley blocks to lift 10 cwts up to 20 tons which cost £3-10s-0d to £14-7s-0d.

Weston Differential pulley blocks for loads of 2 tons to 12 tons cost £3-4s-0d to £19-10s-0d. A whole range of Ships' blocks was made in galvanized malleable steel and a range of monorail electric hoists had been introduced in sizes from 5 cwts up to 5 tons together with electric friction winches for loads of 3 cwts to 20 cwts which cost from £44 up to £90, the latter example being able to lift 20 cwts at a speed of 40 feet per minute to a height of 50 feet and powered by a 5 h.p. electric motor. Forge and foundry cranes were made as self supporting units to cater for loads of 1 to 6 tons and as a crane which could be attached to a suitable upright girder for loads of 1 to 4 tons and costing £38 to £67. Screw jacks and hydraulic jacks also formed part of the Eades Hoist catalogue, these being available in sizes capable of lifting up to 60 tons and included traversing screw jacks which could raise 30 tons and traverse under load up to 18 inches.

Being involved in the motor industry it should be no surprise to find the company making double purpose portable cranes, suitable for lifting and moving motor vehicles at £55 each and also making adjustable engine stands which were comprised of two identical cast iron sides held apart by long threaded steel rods which could be adjusted to the required width and locked into position by nuts on each side of the frames.

In August 1912 plans were submitted to erect a 3 storey building close to the canal which was to house a Machine Shop on the ground floor, an Iron Plate Shop on the first floor and Stores on the second floor. Designed by Ewen Harper & Bros., the building was 116 feet long by 50 feet wide and towered 54 feet high above the canal. The building was divided into two long bays with the ground floor being 24 feet high, the first floor 15 feet high and the second floor 7 feet high. The Machine Shop was obviously intended for very heavy work as two overhead cranes ran the length of the building some 16 feet above the shop floor. The first and second floors were connected via a lift 4 feet 9 inches by 9 feet 6 inches, all three floors also being connected via external staircases. This new building added 17,400 sq.ft. of production space.

Also in 1912, Simon Onions left the Board of Directors and ended the last link between the members of the two families who had come together in 1885 to form this now famous and prosperous company.

DIAGRAM

BLOCK PLAN OF "GREAT WESTERN WORKS" SYDENHAM ROAD - 1916

This block plan shows the works as it appeared in 1916 with many of the original 1888 buildings still in use. The yard and wharf belonging to Robert Coles & Company (civil engineers) later became a coal wharf owned by J.H.King.

- A. Two storey Main Office blocks.
- B. Foundry extension built in 1888.
- C. Core Shop extension built in 1888.
- D. Open sided Shed covering the Travelling Crane, built in 1900.
- E. Three storey building housing a Machine Shop, Iron Plate Shop and Stores. Built in 1912.
- F. Corrugated iron Shed, built in 1914.
- G. Machine Shop, built in 1915.
- H. Existing buildings designated as Engine Shops in 1915.
- J. Two storey Sheet Iron Store, built in 1915.
- K. Existing building designated as a Smiths Shop in 1915.
- e. Entrances into works from Sydenham Road.

PHOTOGRAPHS

LEFT: Both of these photographs were taken in the Great Western Works about 1908 and show Machine Shop Number 1, and Machine Shop Number 3 respectively, both of which appear to have been fully occupied producing engine and other motor car components. All the machines would appear to have been belt driven at this time from the overhead lineshafts clearly visible under the roof trusses.

BELOW: An "Empire" type F, belt driven, positive gas compressor, complete with drive pulley and mounting base which also served as a reservoir. The small diameter pipes supplied water to the jacket around the compressor for water cooling. The largest model in this range had a capacity of 24,100 cubic feet per hour, required a drive pulley 36" in diameter by 10" wide and required a floor space of 5ft 2ins by 8ft 4ins and cost £120 complete.

With the increasing use of coal gas in industry Alldays introduced "Empire" positive pressure gas compressors to enable a uniformly steady pressure, from ½lb to 10lbs/sq.in., to be fed to furnaces, sand blasters and steel converters. These were available in sizes from 5 cu.ft./min up to 660 cu.ft./min and cost from £10 up to £104, both models being belt driven, but models driven directly by electric motors were also available at extra cost. Oil fired furnaces were being added to the long list of articles already made by the company, these being the "Empire" range of portable rivet heating furnaces suitable for ship building etc., and a range of static rivet heating furnaces more suited to workshops. The "Empire" sand mixer and disintegrator was also introduced primarily for use in foundries, but was also suitable for breaking up and/or mixing other materials such as lime, coal, coke, shale, chalk, shells, clay, bones, chemicals, asphalts, cattle foods, soap, rock salt or bark. These were made with reels of 20" diameter up to 36" diameter and with capacities of 1 ton up to 6 tons per hour.

The main offices which ran parallel to Sydenham Road were extended in 1913 by the addition of a first floor building to house four offices. This extension measured 34 feet by 49 feet 6 inches with the offices being 11 feet high and the total height to the ridge being 21 feet. Access to the upper corridor serving the new offices was via a staircase from the ground floor corridor leading to the Telephone Exchange, Strong Room and Mr Alldays' office.

The "Climax" range of Roots blowers had been steadily updated

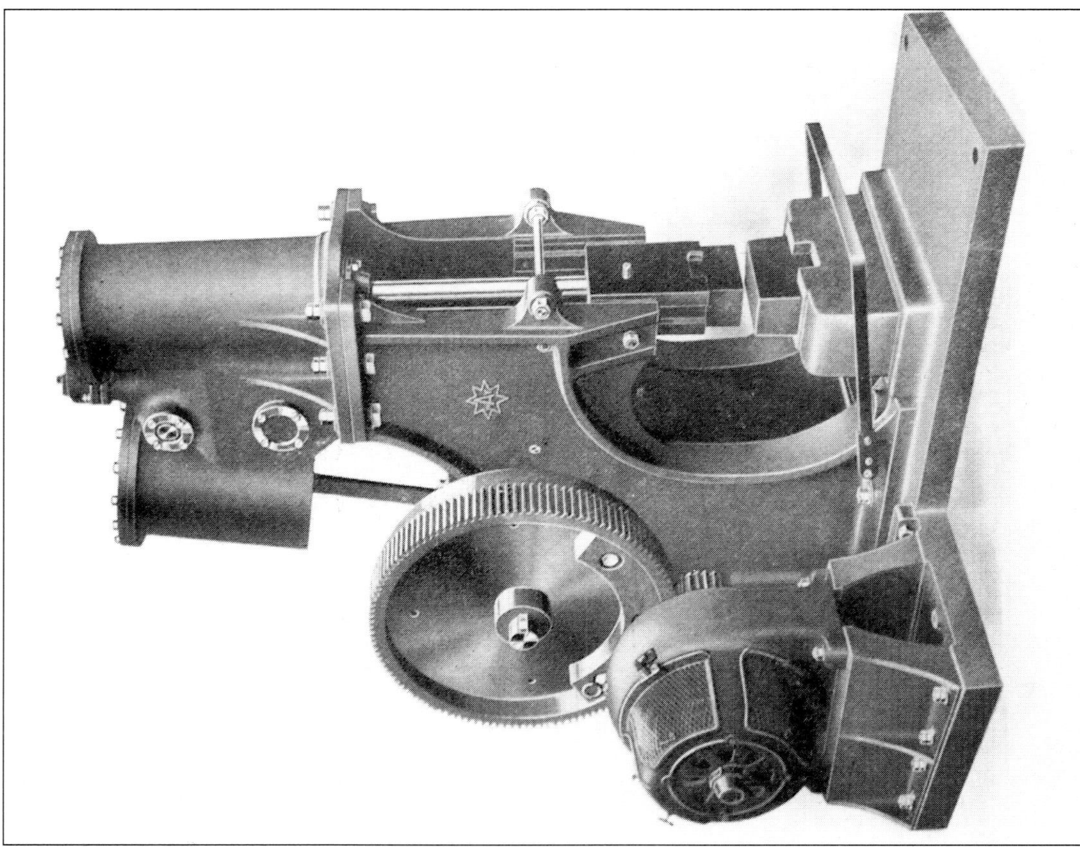

PHOTOGRAPHS

LEFT-TOP: The Alldays improved Spring Type Double Oliver, a small foot operated hammer for producing small forgings such as bolts, collared pins and cutlery.

Operated by the foot operated treadle at floor level, the two hammer heads could be adjusted sideways to enable two operations to be undertaken together where a roughly shaped job could be done on one side of the anvil block and then finally finished off by the second hammer. The machine weighed a few pounds over one ton and cost £22.

LEFT-BOTTOM: A 10 cwts Series 'B' motor driven pneumatic forging hammer for much heavier forging work which could draw out a 10" diameter steel bar to 2½" by 1½" at one heating of the bar. This size of hammer would be used for forging such items as a 14lbs axle, 28lbs bench vice body or 9lbs hook for lifting tackle.

The complete hammer as shown here weighed 12½ tons, stood 11ft 6ins high and could deliver up to 110 blows per minute. The cost was £493.

ILLUSTRATIONS

RIGHT: What could be termed miscellaneous items made by the company, but important nevertheless to the foundry industry.

The coke barrow with a body measuring 30" by 18" cost £1-10s-6d.

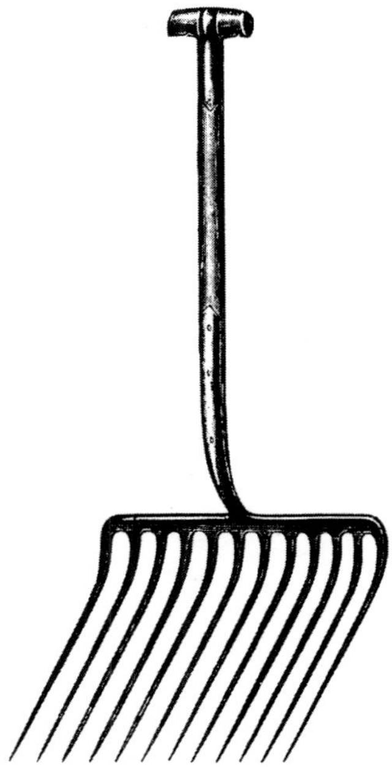

PATTERN NO. **1650**

COKE FORKS,
With Strapped Handles.

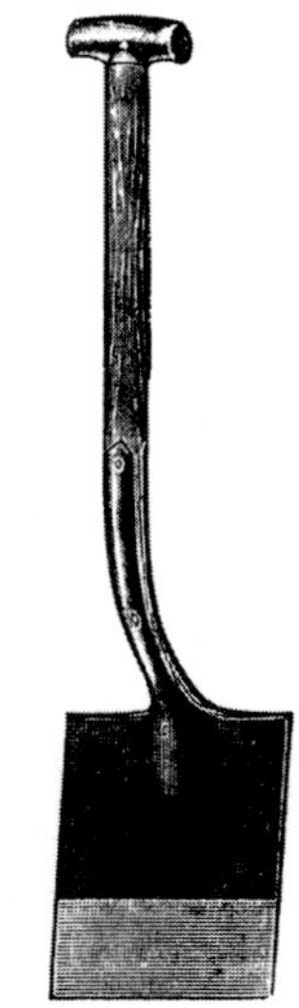

PATTERN NO. **1680**

MOULDERS' SPADES,
Hammered and Strapped.

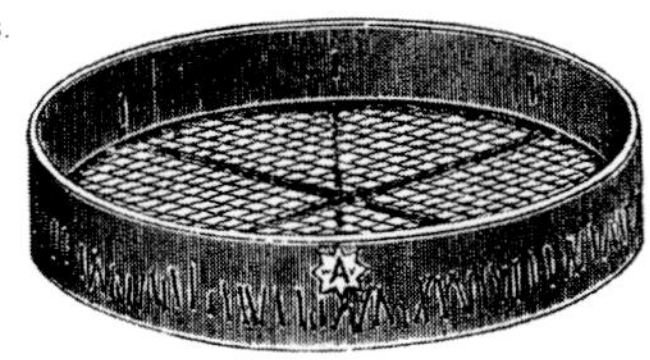

PATTERN NO. **10 F.R.**

FOUNDRY RIDDLES AND SIEVES,
With strong wooden rims, and perfect and strong woven mesh. Of special finish for foundry use. Made in all sizes from 16in. to 24in. in diameter.

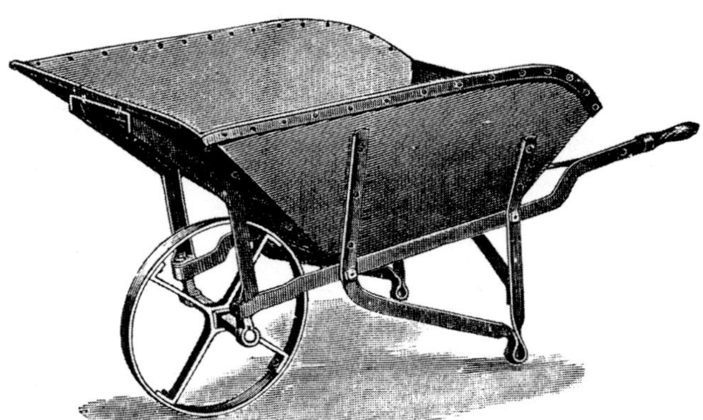

PATTERN NO. **1660**

WROUGHT-IRON COKE BARROWS,
in all sizes.

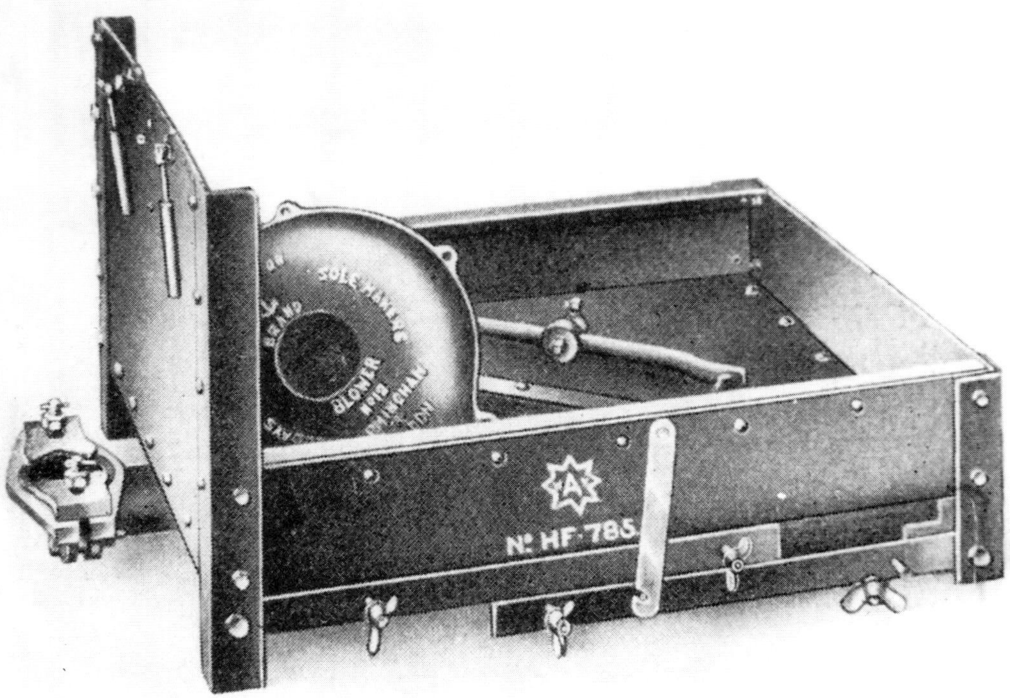

The "MOBILE" Forge folded for transport.

PHOTOGRAPHS

LEFT: A typical scene in a British Army engineering workshop with an Alldays "Mobile" forge in use. In all probability the portable vice bench to the left and the anvil to the right were also Alldays products.

The "Mobile" forge itself is shown in the lower picture with its legs folded up and the hand operated blower removed all ready for transport.

BELOW: A new pattern Number 3 size "Climax" roots blower powered by a single crank 'open type' steam engine.

'Open type' steam engines were made in various sizes which included 4½"x4", 6½"x 5½", 9"x6", 10"x9", 13"x9" and 15"x12".

"Enclosed" engines, where the con-rod and crankshaft were totally enclosed within a cast iron casing were made in a much larger range of sizes which included 3"x3½", 4½"x3½", 6"x4", 7½"x 5", 9"x15", 10"x16", 12"x16", 14"x7½", 14"x9" and 17½"x9".

The blower and engine illustrated below produced 3,000 cu.ft. of air per minute at 380 r.p.m. and the engine developed a maximum of 12 b.h.p. This required a floor space of 54" by 80", weighed 1 ton 17 cwts complete and cost £140.

and improved and now included models directly driven by electric motors, although belt driven and steam driven models were still available, the latter now available with Pickering steam governors. A "Climax" automatic speed regulator was also introduced to give control of variable speed motors driving blowers, enabling the blower speed to be reduced automatically as the number of hearths being fed shut down instead of the excess air being released through an escape valve whilst the blower was still being driven at full speed.

Just twenty days after the start of the First World War plans were submitted for a large corrugated steel shed to be erected parallel to the canal bank, the building measuring 60 feet by 18 feet stood 12 feet high to the eaves and 18 feet 3 inches high to the ridge, but no indication was given as to its use on the company's own drawing, number 6960.

The company designed and built its own shell nosing furnace and added this to the product list and, although the military tank was just in its infancy, Alldays also designed and built a special riveting machine used to construct the tracks for tanks and this was offered, after the war, for civilian applications.

Like many Birmingham engineering companies at this time Alldays were set to work producing parts for shells and, to this end, installed additional drop stamping hammers for their production. The company's portable forges were in great demand at this time for the armed forces and these included the "Hurricane" and "Typhoon" hand geared fan forges, the "Empire" lever fan forge and the "Mobile" forge which could be folded up for transporting and became a useful addition to many army workshops in the field.

Alldays Improved "Midget" Portable Fan Forge and Case.

No. A 604.

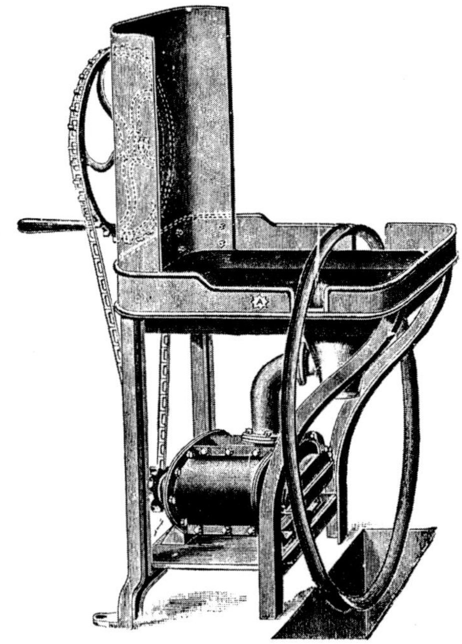

A340

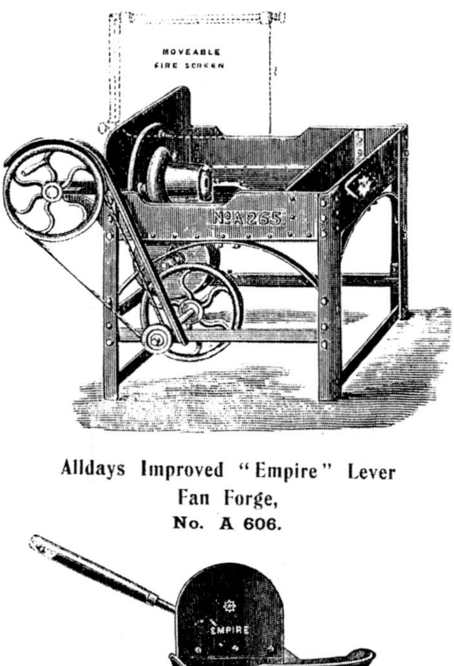

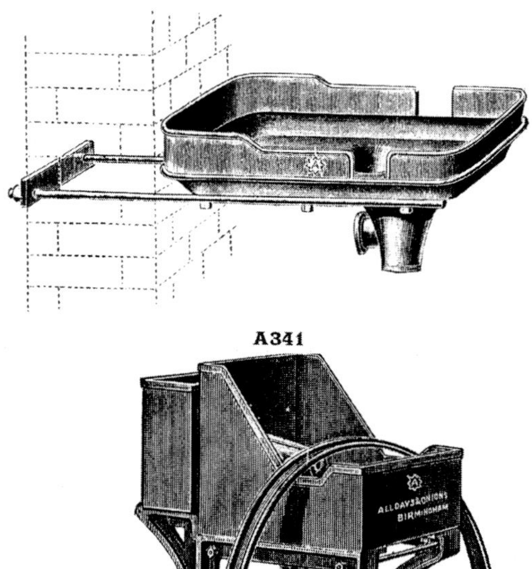

A341

Alldays Improved "Empire" Lever Fan Forge,

No. A 606.

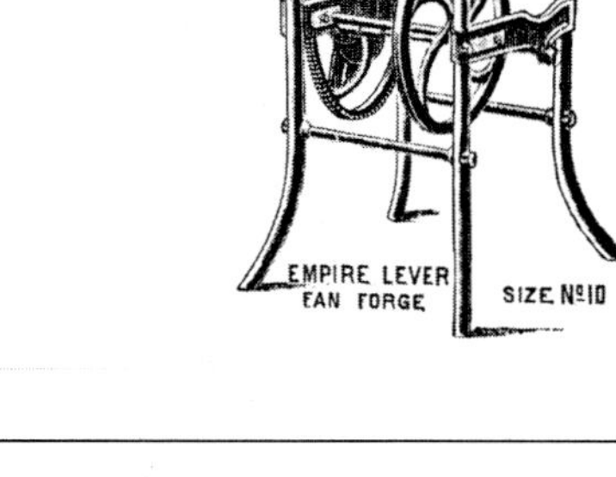

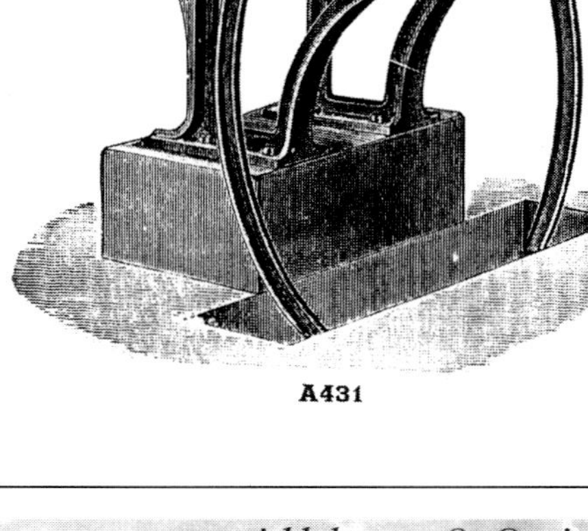

A431

ILLUSTRATIONS - LEFT

LEFT: The left hand illustrations, from top to bottom, show the following:

An improved "Midget" portable fan forge constructed from cast iron with a hand operated fan driven via a toothed wheel and pinion. The working pan measured 12" by 17" and the overall height of the forge was 32". For transporting, the forge packed away into the wooden case shown here which measured 22½" by 18½" by 16½".

The "Branston" portable fan forge also featured a hand operated fan and was made in three sizes, with the working pans measuring 20" by 26", 24" by 30", or 34" by 30". These all weighed about 2 cwts and could heat metal bars between 1½" diameter and 2" diameter.

The "Empire" lever fan forge was made from cast iron and had a working pan 20" by 16" and stood 30" high.

The illustrations on the right hand side show, from the top, a "Climax" forge fitted with a hand operated Roots blower and made for heat welding steel tyre rims used for wheels fitted with rubber tyres; a wall mounted version of this forge, but with air blast supplied by a remote blower, and a coach builder's hearth with water cooled tue iron also made for heat welding of steel tyre rims. The steel tyre rims made on these hearths were of the type used by carriage builders, and early motor cars, which used solid rubber tyres.

ILLUSTRATIONS

RIGHT-TOP: Examples of the split wrought iron pulleys featuring a single row of arms between the hub and rim on the left and double row of arms on the right which were recommended when the width of the pulley drive face was 12" or above. These pulleys were made in a range from 10" diameter by 4" wide up to 100" diameter by 12" wide.

RIGHT-BOTTOM: A "Paragon" Number 3 emery grinding machine of a type awarded the Silver Medal at the Paris Exhibition of 1889. This heavy duty machine featured a main drive spindle 2.75" diameter and was available with emery wheels 24" diameter 3" wide one side and 1½" wide the other, 20" diameter with 2½" wide and 1½" wide wheels and with wheels 18" diameter by 2" wide and 1½" wide. These machines cost £30, £24 and £21-10s-0d respectively.

The largest grinding machine listed in an Alldays catalogue was the "Paragon" Pattern 5850 type, complete with water trough, which had a grindstone wheel 72" in diameter by 12" wide, was driven at 55 r.p.m. from a belt pulley 30" in diameter by 8" wide, and cost £70 in 1900. The operator of this monster machine, which weighed just over 3 tons, sat astride the grinding wheel on an oak "horse", or saddle, supported by adjustable 'risers' to enable the "horse" to be kept as close as possible to the wheel as it wore down with use.

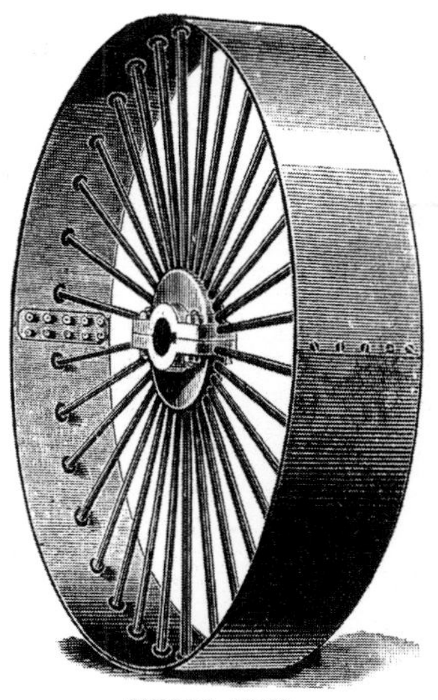

SINGLE ARMS.

DOUBLE ARMS.

Silver Medal, Paris 1889.

"Paragon"

Universal Emery Grinding Machine.

No. 3.

PHOTOGRAPHS

LEFT-TOP: A Smithy's Shop fully equipped with Alldays products, which include hearths, blower for supplying the air blast to the hearths, overhead extraction equipment, grindstone and all the necessary tools.

LEFT-BOTTOM: A brass foundry equipped with Alldays fume extraction equipment.

BELOW: A New Pattern Number 10 size "Climax" Roots Blower complete with 80 h.p. AC electric motor mounted on a single bedplate. Electrically driven blowers were by this time rapidly taking over from the old steam powered models, making much simpler installation possible. This was the largest Roots blower then offered by the company and produced 19,400 cu.ft. of air per minute at 245 r.p.m. and could feed up to 220 blacksmiths' fires. This required a floor space of 18ft by 6ft 9ins, weighed 13.8 tons without the electric motor and cost £390.

A serious fire in April 1915 swept through the Packing Case Department in Sydenham Road which by this time was being used exclusively for the manufacture of wood drums used to transport barbed wire for the army.

In May 1915 planning permission was sought for a new Machine Shop at the Sydenham Road works. The new building, which measured 120 feet by 90 feet, was built on the site of some of the original factory workshops and was flanked by two buildings designated as Engine Shops. With a height of 20 feet 9 inches to the eaves and 34 feet 6 inches to the ridge it had a clear centre span which was 25 feet high and beneath the building was a basement about 8 feet high. (This basement became an air raid shelter in World War Two with two sloping ramps going down into it from the workshop above and still formed part of the factory in 1998 when the works was owned by Wallwork Ltd., who specialised in hardening and heat treatment processes.) Work on the new machine shop was completed by the last day of 1915 and, completed at the same time was a small two storey extension to buildings at the Golden Hillock Road end of Sydenham Road on the perimeter of the Great Western Works, to house a Sheet Iron Store on the ground floor and a Drying Room on the upper floor. Measuring 44 feet 9 inches long by 24 feet wide this extension was built to match the existing buildings and was 22 feet high to the eaves, 30 feet high to the ridge with the ground floor having a height of 10 feet 6 inches and the upper floor being 11 feet 6 inches high. This building also contained a basement about 6 feet high. The architects engaged to design these buildings were Ewen Harper, Brother & Co.

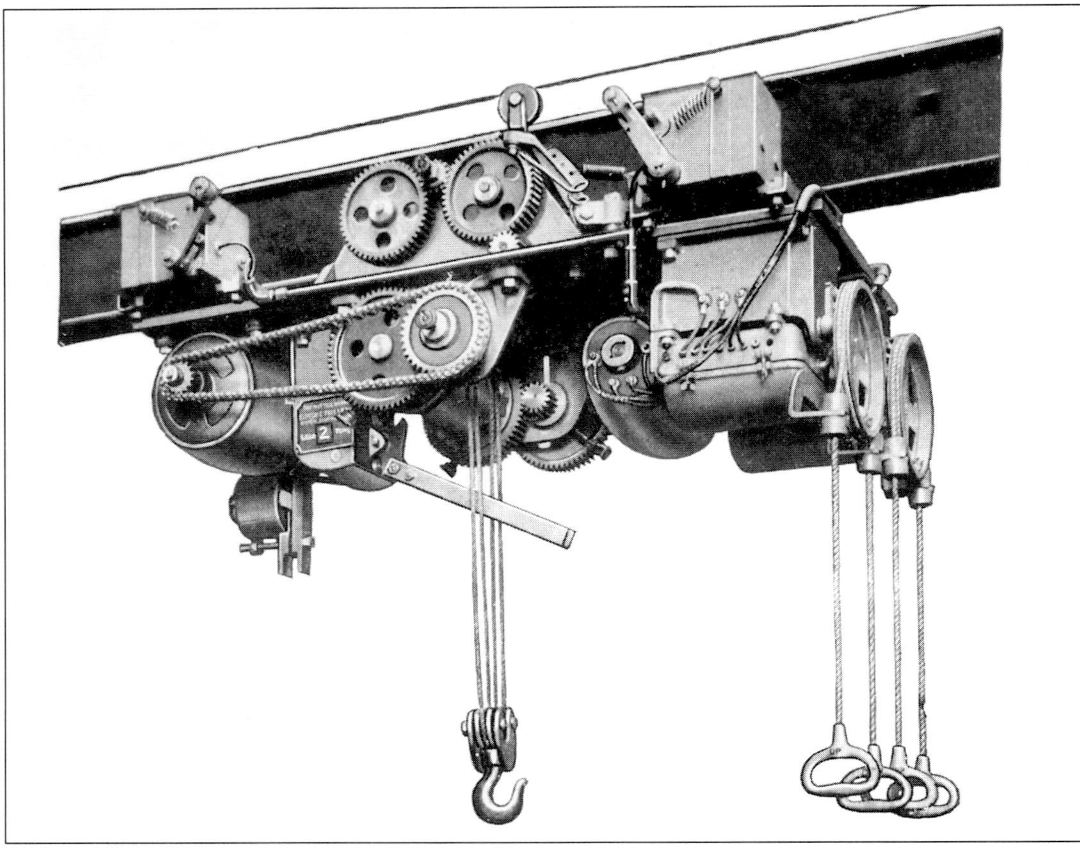

CONTRACTORS TO THE ADMIRALTY, THE WAR OFFICE, THE INDIA OFFICE, THE POST OFFICE, COLONIAL & FOREIGN GOVERNMENT, THE CROWN AGENTS FOR THE COLONIES.

ALLDAYS & ONIONS, PNEUMATIC ENGINEERING CO. LTD

ESTABLISHED 1650.

MANUFACTURERS OF

SMITHY & FOUNDRY PLANT. COMPLETE HARDENING SHOP EQUIPMENT. VENTILATING APPLIANCES. ETC.

REG. TRADE MARK.

LONDON OFFICE:- 58. HOLBORN VIADUCT, E.C.1.
TELEPHONE: 5948 HOLBORN.
TELEGRAMS & CABLES:- "TYPHOON, CENT. LONDON."

GREAT WESTERN WORKS.
BIRMINGHAM.

TELEPHONE:- VICTORIA 487. PRIVATE EXCHANGE.
TELEGRAMS & CABLES:- "ALLDAYS. BIRMINGHAM.
CABLES: A.B.C. 4TH & 5TH EDITION.

SAS/MC

Your Ref.

Our Ref.

September 16th.1918.

MUNITIONS OF WAR ACT 1915 CONTROLLED ESTABLISHMENT

PHOTOGRAPHS

LEFT-TOP: Although made under the Eades Hoist Co. Ltd. name all this lifting equipment was manufactured in the Great Western Works.

This picture shows a 2 ton capacity electrically powered travelling hoist which ran on a monorail.These were made in sizes from 5 cwts up to 5 tons and could be supplied with all the necessary runways and junctions to operate the hoist throughout a number of workshops, etc. The hoist itself featured an electro-magnetic brake mechanism.

LEFT-BOTTOM: The Alldays Double Purpose Portable Crane designed specially for use in motor car garages and capable of lifting 30 cwts. The crane weighed 10 cwts and cost £55.

RIGHT: Being so involved with the motor industry itself it is not surprising to find the company making these adjustable engine stands which cost £3-10s-0d. The stand is made from two identical cast side frames which are separated by adjustable threaded rods, enabling the stand to adjust to a large number of different engine widths. Each leg was provided with a small wheel allowing the whole thing to be moved easily about and each wheel was provided with a brake lever to prevent movement when working on the engine.

(SINGLE INLET.)

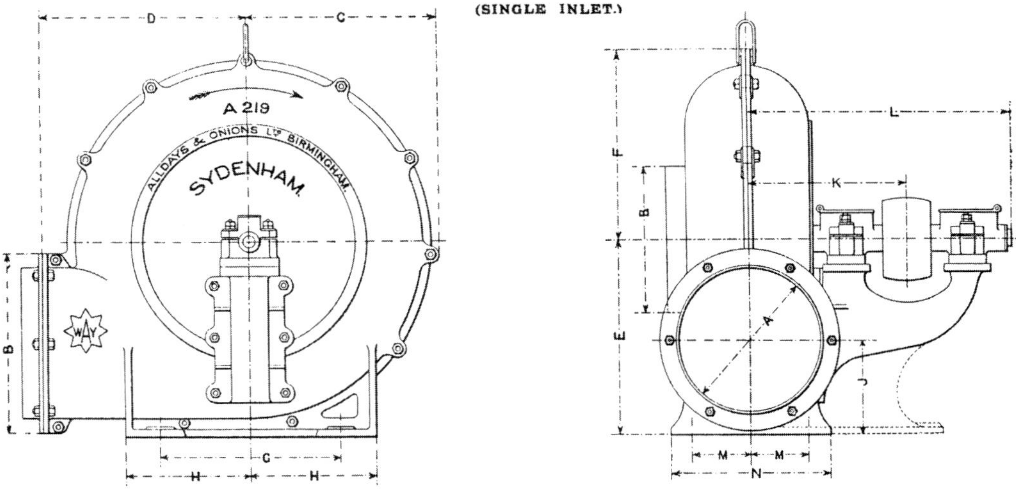

(SINGLE INLET TYPE.)

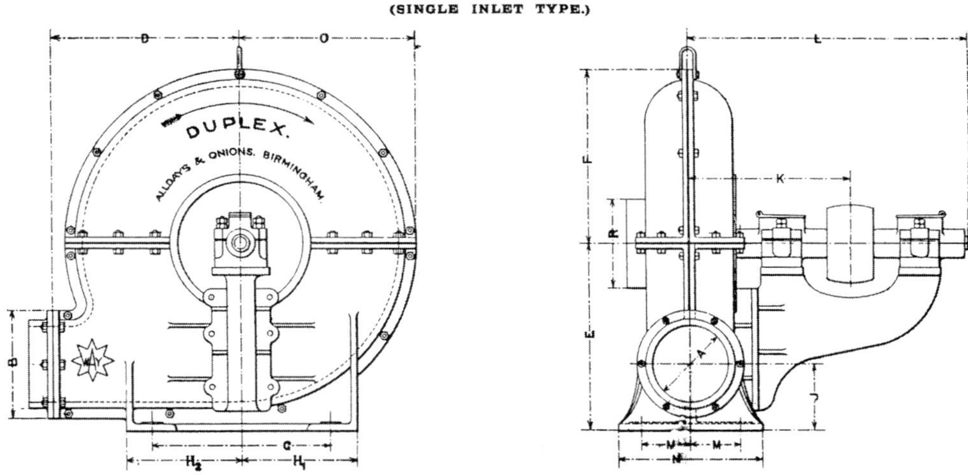

(DOUBLE INLET TYPE.)

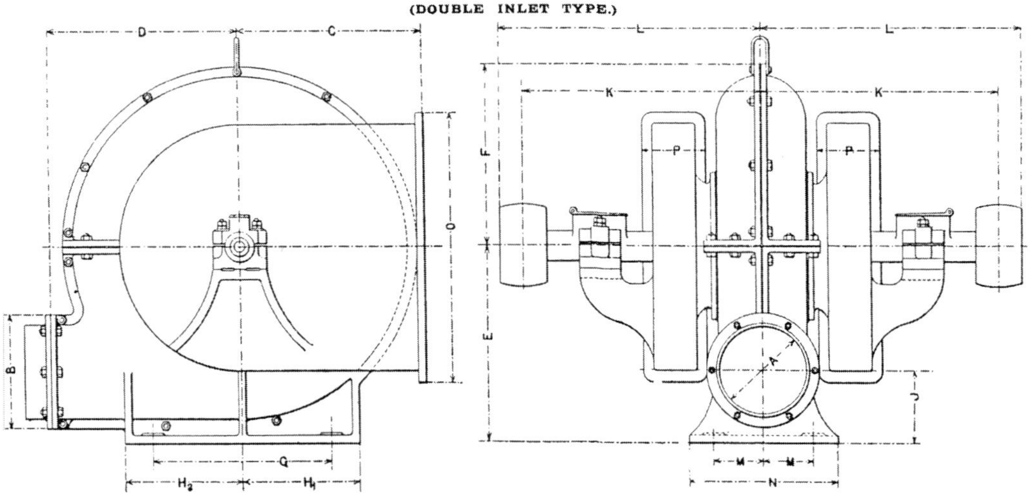

ILLUSTRATIONS

LEFT: Drawings of "Sydenham" and "Duplex" exhaust fans which featured cast iron casings made in two halves in the "Sydenham" design and in four parts in the "Duplex" design.

The top drawing is of a single inlet fan with belt drive to the right hand side and ranging in size from a special size 9 which stood 14.5ins high, 15.75ins long and weighed 42lbs up to the size 50 which stood 64ins high and long and weighed 1 ton 9 cwts.

The centre drawing is of a single inlet "Duplex" fan, again with single belt drive to the right hand side. The difference in the casing construction can readily be seen here with the casing split on the vertical and horizontal centre lines. The smallest fan in this range was the size 10.25 which stood 14.75ins high and 14.5ins wide. The largest, size 104.25, stood 122.5ins high and wide with the pulley 24ins in diameter by 12ins wide.

The bottom illustration shows a "Duplex" double inlet fan with twin belt drives. The overall dimensions of the largest fan in this range were 122.5ins high, 123 ins long by 132ins across the width of the pulley drives. The two inlets of this fan each measured 45ins by 24ins with the single outlet 30.25ins in diameter. This fan had a capacity of 20,274 cu.ft. per minute at 180 r.p.m. and cost £255.

RIGHT: This illustration shows the "Serlano" wall fan which was made in a range from 12ins in diameter up to 48ins in diameter and with capacities from 625 cu.ft. per min up to 10,000 cu.ft. per min. These were designed to bolt directly into the wall of a workshop and were driven by a single belt pulley. The largest fan in this range had a cast iron frame 56ins square, could be fitted into walls 18 to 22 inches thick, was driven from a pulley 13.5ins diameter by 5ins wide and weighed just over 14 cwts.

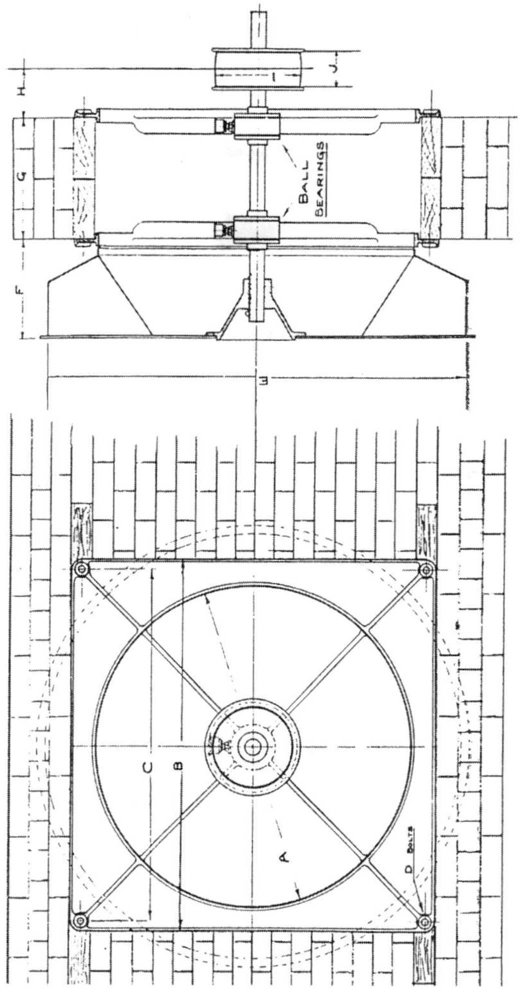

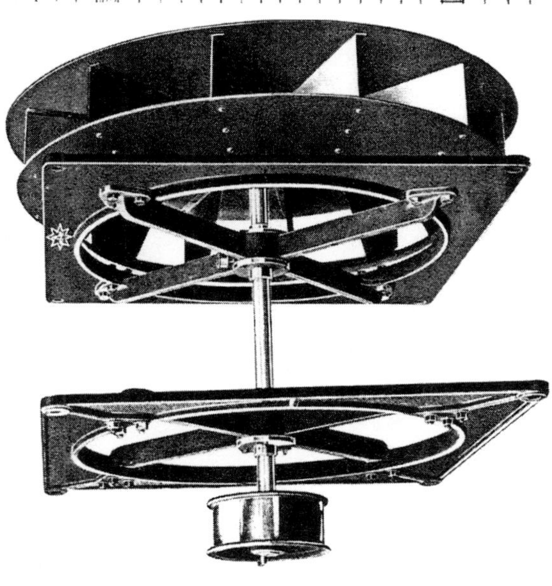

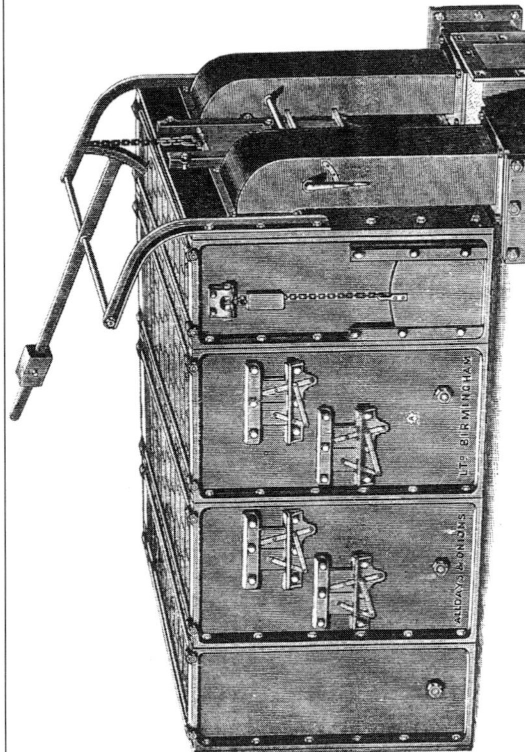

Specially Recommended for Spring Makers, Steel Toy Works, Pen Makers, Gun Work, Cycle Works, &c.

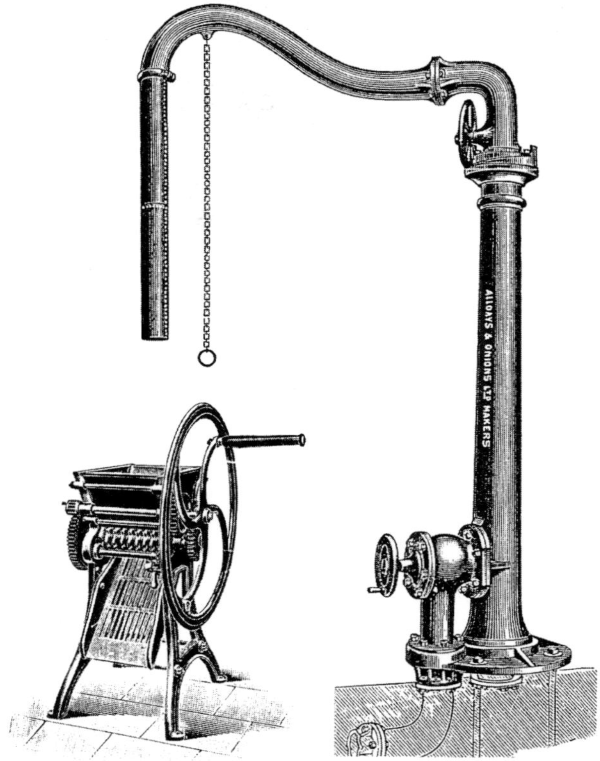

PATTERN NO. **20 C.B.**

IMPROVED COKE BREAKERS,

PATTERN NO. **50 D.F.L.**

UNBREAKABLE DOUBLE-HANDED FOUNDRY LADLES.

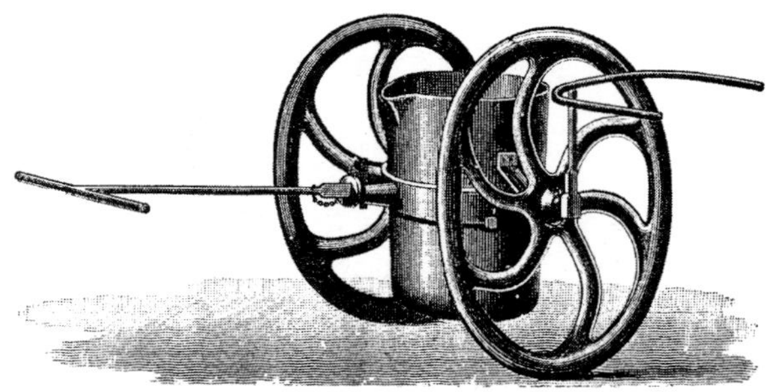

PATTERN NO. **15 B.L.**

BOGEY LADLES,
From 6 cwts. to 12 cwts. capacity.

ILLUSTRATIONS

LEFT-TOP: Several kinds of furnaces were made like this large Spring Heating furnace with counterbalanced door and the small Case Hardening Furnace for the heat treatment of components in the cycle and small arms industries.

The photograph shows two "Empire" coal fired furnaces complete with Babcock boilers heated with the waste heat from generating steam for power hammers installed in a large railway works.

RIGHT: Miscellaneous items of Alldays huge range of products, showing from the top, a water tower used for filling railway locomotive boilers, a hand operated coke breaker for smithy's work and for the foundry, a double handed ladle made in stamped steel or wrought iron to hold 84lbs up to 896lbs and a wheeled bogey ladle for handling molten metal and made in a range capable of carrying 6 cwts up to 12 cwts.

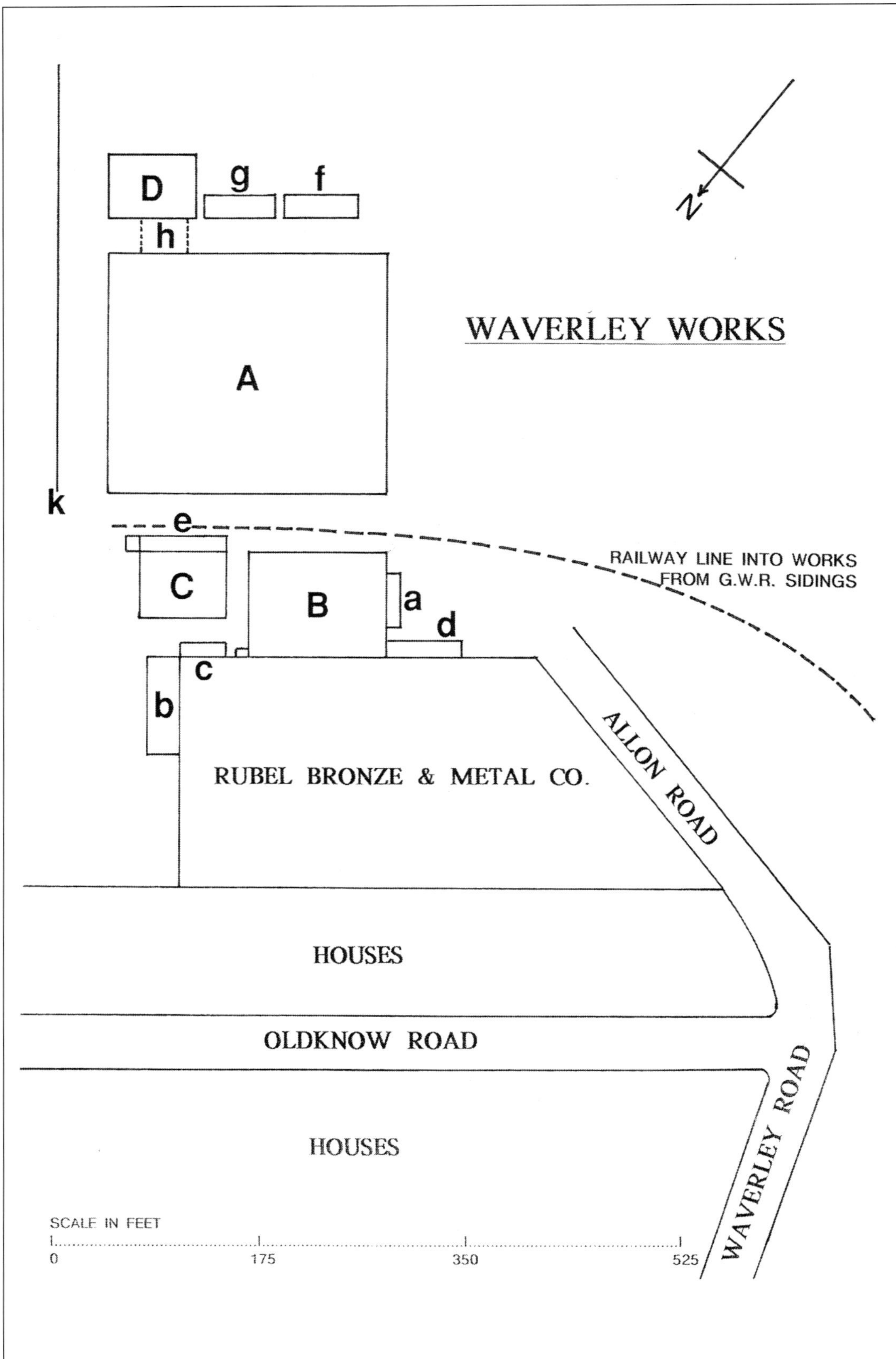
D
g
f
N
h
WAVERLEY WORKS
A
k
e
RAILWAY LINE INTO WORKS FROM G.W.R. SIDINGS
C
B
a
d
c
b
RUBEL BRONZE & METAL CO.
ALLON ROAD
HOUSES
OLDKNOW ROAD
WAVERLEY ROAD
HOUSES
SCALE IN FEET
0
175
350
525

DIAGRAM

BLOCK PLAN OF THE "WAVERLEY WORKS" 1922.

- A. Body Building Shop & Mounting Shop 240 feet by 200 feet.
- B. Sawmill 120 feet by 90 feet.
- Ba. Engine House attached to Sawmill 42 feet by 19 feet 6 inches.
- b. Timber Store 85 feet by 28 feet.
- C. Temporary Pattern Shop 60 feet by 73 feet.
- c. Shed for small castings 42 feet by 10 feet 6 inches.
- Ce. Platform 84 feet long by 12 feet wide with ramp at one end. Used to load/unload railway trucks.
- D. Plate Shop & Iron Shop 74 feet by 54 feet.
- d. Storage Shed for bellows boards 63 feet by 14 feet.
- f. Engine Test House 61 feet 6 inches by 21 feet.
- g. Temporary Hardening Shop 60 feet by 20 feet.
- h. Temporary Car Washing Shed 40 feet by 28 feet.
- k. Retaining wall built to cater for the changes in site levels.

THE WAVERLEY WORKS

A large area of land known as the Waverley Estate to the rear of a row of houses in Oldknow Road, Small Heath, and bordered by the Grand Union Canal and River Cole, was being developed for industrial use and Alldays applied for planning permission in September 1914 to build a factory there to cater for the increasing numbers of motor vehicles being built. This must have been an enormous investment for the company which was obviously already finding problems in trying to increase production on the Fallows Road site where practically every inch of space was already being used and the Sydenham Road factory was fully employed with general engineering work and providing engine, gearbox and axle castings for motor vehicle production.

The first building to be erected was a single storey workshop measuring 240 feet long by 200 feet wide and built in 12 bays, each being 20 feet wide, with a height of 12ft 9 ins to the eaves and 21 feet high to the ridges. This building alone covered just over one acre and was used as a motor car body building shop and body mounting shop for chassis built at the Matchless Works.

The second building to be erected was a Sawmill 120 feet long by 90 feet wide located 50 feet away from the first workshop. Using wooden 'Belfast' roof trusses the sawmill had a single clear span of 90 feet and was 14 feet high to the eaves and 23 feet high to the top of the curved roof, plus a further 4 feet to the top of a central ventilator. Work on this building commenced on January 3rd 1916 and was completed by October 9th, 1916. During its construction an Engine House to supply the new factory buildings with electricty was added to the sawmill measuring 42 feet long by 19 feet 6 inches wide and 9 feet high to the eaves and 12 feet 9 inches high to the ridge, the engine house being completed by June 22nd 1916. The architect engaged to design these buildings was Ewen Harper, Brother & Co. In September 1918 a Timber Store was added to the Waverley Works close to the Sawmill and built at the corner of a new factory building belonging to the Ruben Bronze & Metal Co. This open fronted building measured 85 feet by 28 feet and was 16 feet high to the eaves and 24 feet high to the ridge and was used to store timber for the sawmill where all the various parts were machined which made up motor car and possibly commercial vehicle bodies. During the early months of 1918 three large sheds 60 feet long and with a total width of 43 feet had also been erected close to the Sawmill which were used as a Pattern Shop and permission was sought to erect two more similar sheds, each 15 feet wide, to expand this department. A letter to Birmingham City Council Surveyors Department expressed the urgency attached to these buildings, part of which read:- "You will be aware from our previous correspondence that these buildings are required to carry out work which we have in hand for the Mechanical Welfare Department in London, for whom we hold a large number of contracts carrying a very high priority, Class "A" P1, and as the work is of greatest national importance and of considerable urgency we shall be glad if you will be good enough to sanction this extension with as little delay as possible." This extension was granted, but the buildings had a temporary licence for 3 years only, like many factory extensions built during the First World War. A single railway track had been laid into the Waverley Works from the nearby Great

Alldays & Onions Ltd
Birmingham & London

Alldays & Onions Ltd
Birmingham & London

PHOTOGRAPHS

LEFT: Both of these vehicles built in 1916 would have had their bodies built and mounted onto their chassis in the Waverley Works. The top picture shows a two seater "Light Four Coupé" which cost £255 complete with folding leather hood, five lamps, horn and detachable steel wheels and upholstered in either Bedford Cord or leather.

The lower picture shows a 5 cwts delivery van based on a "Light Four" chassis frame and cost £170 complete with two side lamps, tail lamp, horn, detachable steel wheels with spare, and painted in Alldays dark green. However, with the windscreen shown here, it cost an extra £3-3s-0d!!

BELOW: This picture, taken in 1917, shows an Alldays 2 tonner in service with Kynoch Ltd., and used for carrying all kinds of munitions from their works in Witton, Birmingham. Kynoch had also put a 5 tonner into service in 1914. The cab and dropside body for all the company's commercial vehicles would also have been made in the Waverley Works.

(Photograph courtesy of Birmingham Library Services)

Western Railway which then changed into two tracks where it crossed Allon Road, the top of the tracks being level with the road surface. Alongside the buildings making up the Pattern Shop was a raised platform 84 feet long by 12ft 6ins wide with a ramp at one end. Clearly the railway was being used by the company to bring in raw materials and possibly to despatch completed vehicles and other wartime products.

A roadway from the junction of Waverley Road and Oldknow Road was built to access the new site and by 1918 this had become known as Allon Road, the name being derived from Alldays & Onions!

As post war production resumed the Enfield Autocar Company name was discontinued and the name Enfield-Allday Motors adopted with all production of light cars concentrated at the Waverley Works. The only Enfield-Allday commercial vehicle to be listed as in production at this time was a 2 tonner powered by a 25/30 h.p. 4 cylinder engine with a bore of 100mm diameter and stroke of 130mm, with the complete vehicle costing a hefty £852, compared to the pre-war cost of £530. Post war the market for new commercial vehicles had become difficult to sustain, due to the influx of ex.W.D. vehicles coming back onto the market.

In February 1919 a Plate Shop & Iron Shop measuring 54 feet by 74 feet was erected to the rear of the Body Shop leaving a roadway 30 feet wide between these buildings. In September of the same year a temporary Car Washing Shed was erected against the rear wall of the Body Shop, this open ended structure in two bays being 40 feet long by 28 feet wide, 15 feet high to the eaves and 20 feet high to the ridges. The need for this shed would indicate that motor cars were being road tested in chassis form before going to the Body Shop for the bodywork to be mounted onto the chassis and that the running chassis would be washed down before entering the mounting area of the Body Shop.

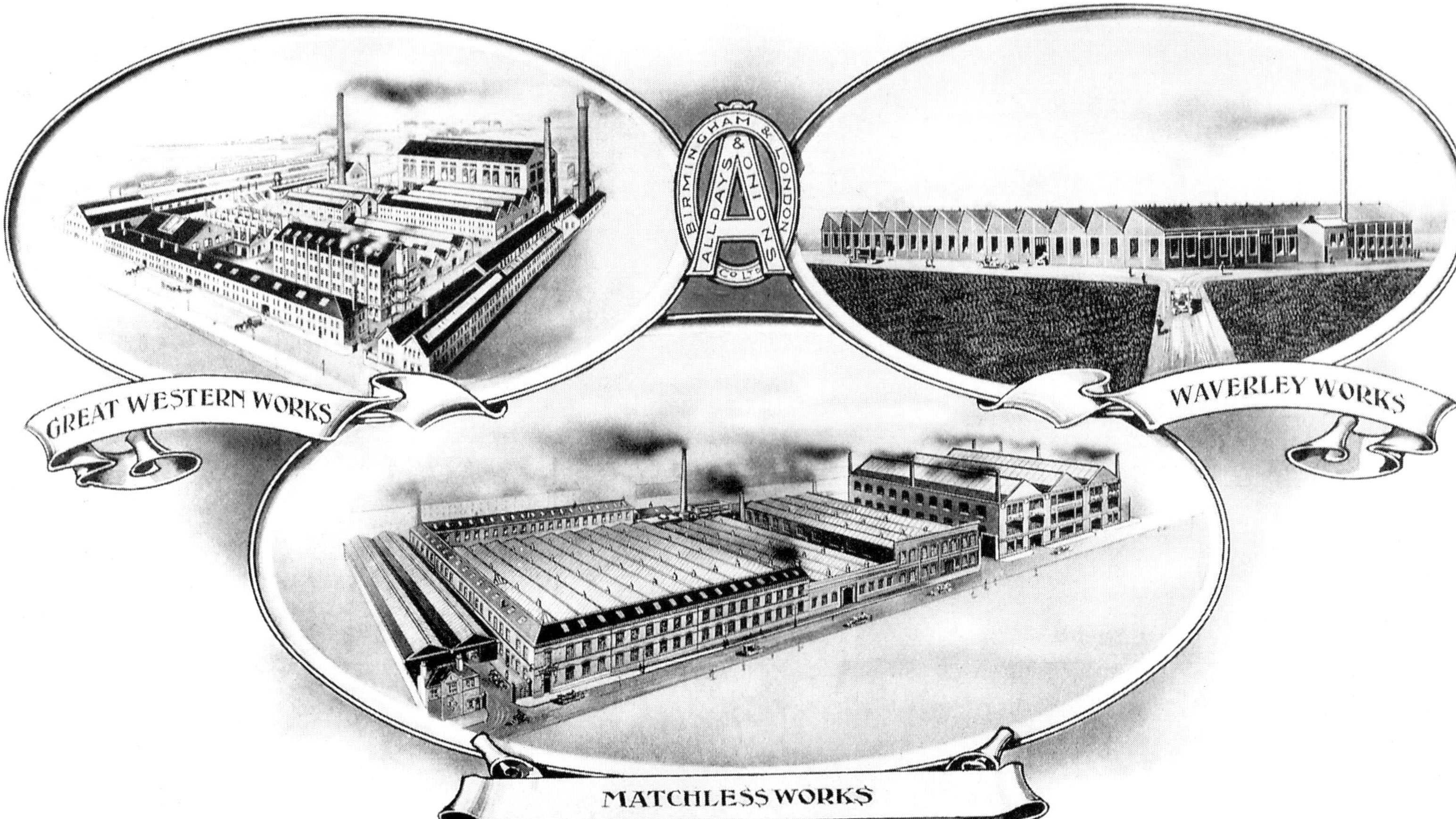

OUR GREAT WESTERN ENGINEERING WORKS are equipped with all the latest machinery for the manufacture of General Engineering Goods and numerous specialities which are illustrated in a special list.

OUR MATCHLESS WORKS, where our "ALLON" Motor Cycles and Motor Cars are made, have recently been extended and fitted up with all the latest tools and appliances suited to our manufactures, and the above illustration only gives a faint idea of their size and the ground on which they stand.

WAVERLEY WORKS.—This new extension is where our Body Building and Coach Departments are carried out, for Commercial Vehicles and Pleasure Motor Cars.

ILLUSTRATIONS

LEFT: This montage of factory pictures shows an artist's impression of how the three factories looked in 1918. Apart from two buildings being added and the horse drawn carts being replaced by motor vehicles the Great Western Works picture is the same as one being used in 1900. The Waverley Works picture shows the Body Building Shop and Mounting Shop, building 'A' on the block plan, but the Engine House shown to the right of this building was a figment of the artist's imagination. The Engine House for the whole site was built onto the side of the Sawmill, building 'Ba' on the block plan! However, the view of the Matchless Works is reasonably accurate, the only errors being in the construction of the roofs on two of the buildings.

BELOW: A 1938 map of part of Small Heath showing how the Waverley Works site had been extended, but, by this time of course, much of the site was owned by B.S.A. for production of bicycles. Alldays & Onions had vacated their motor car works on the site in 1923.

In March 1921 a small Weighbridge Office was built measuring 20 feet by 9 feet 6 inches connected to a 30 ton weighbridge built into the single track section of railway line. This building was 9 feet 6 inches high to the eaves and 12 feet high to the ridge. In November of the same year an Engine Test House was also built on this site to the rear of the main assembly building. This building measured 61 feet 6 inches long by 21 feet wide, 13 feet high to the eaves and 18 feet high to the ridge. This was an interesting development indicating, possibly, that engines were being assembled on this site as well as in the Great Western Works.

The final building to be erected at the Waverley Works, in 1922, was a temporary Hardening Shop made of corrugated steel sheeting and measuring 60 feet long by 20 feet wide, 13 feet high to the eaves and 18 feet 8 inches high to the ridge. This was located next to the Engine Test House to the rear of the main assembly building.

In the latter part of 1923 the company was facing increasing financial difficulties, mainly through losses associated with the motor car side of the business, and the Waverley Works was closed down. This pattern of rapid expansion before and during the First World War followed by post war financial difficulties was experienced by many companies associated with the motor car industry in particular and led to the complete collapse of companies much larger than Alldays & Onions. However, Alldays withdrew into the Great Western Works under the direction of the Receiver Mr.Herbert Smith and was reconstituted in July 1925 as the New Alldays & Onions Ltd. The company was then in the ownership of Sir William Mitchell Cotts and Mr. A.T.Cocking J.P.

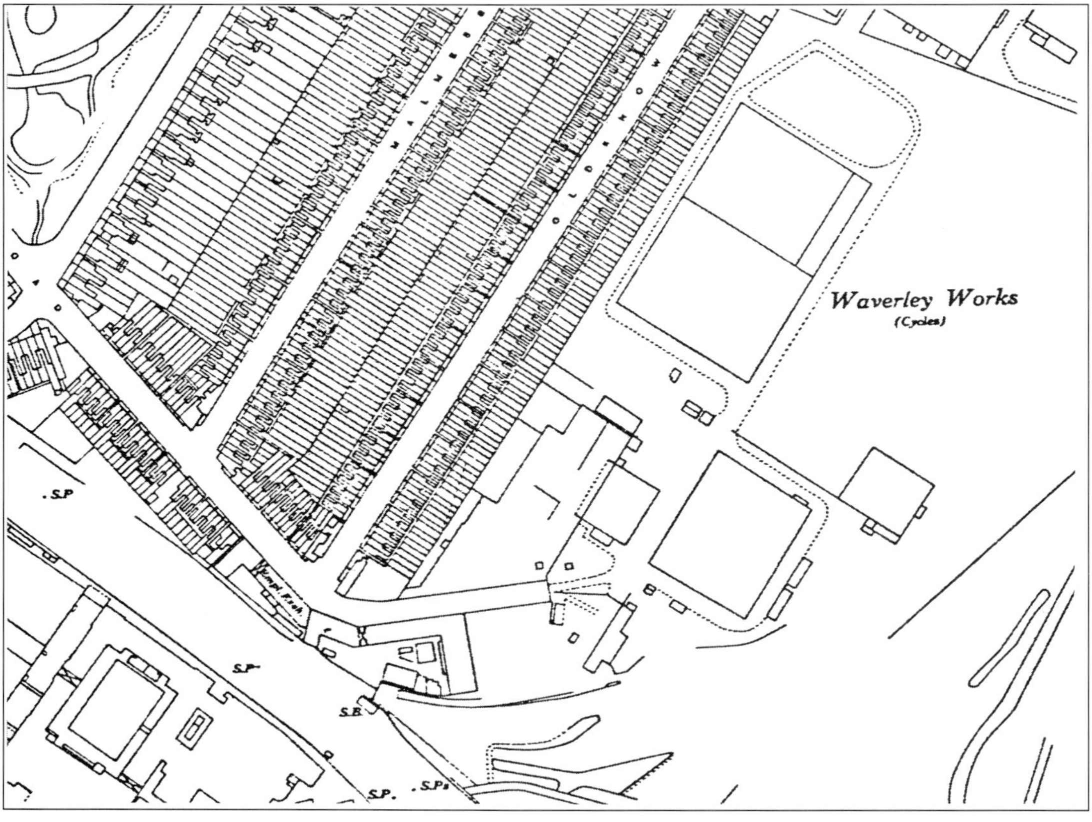

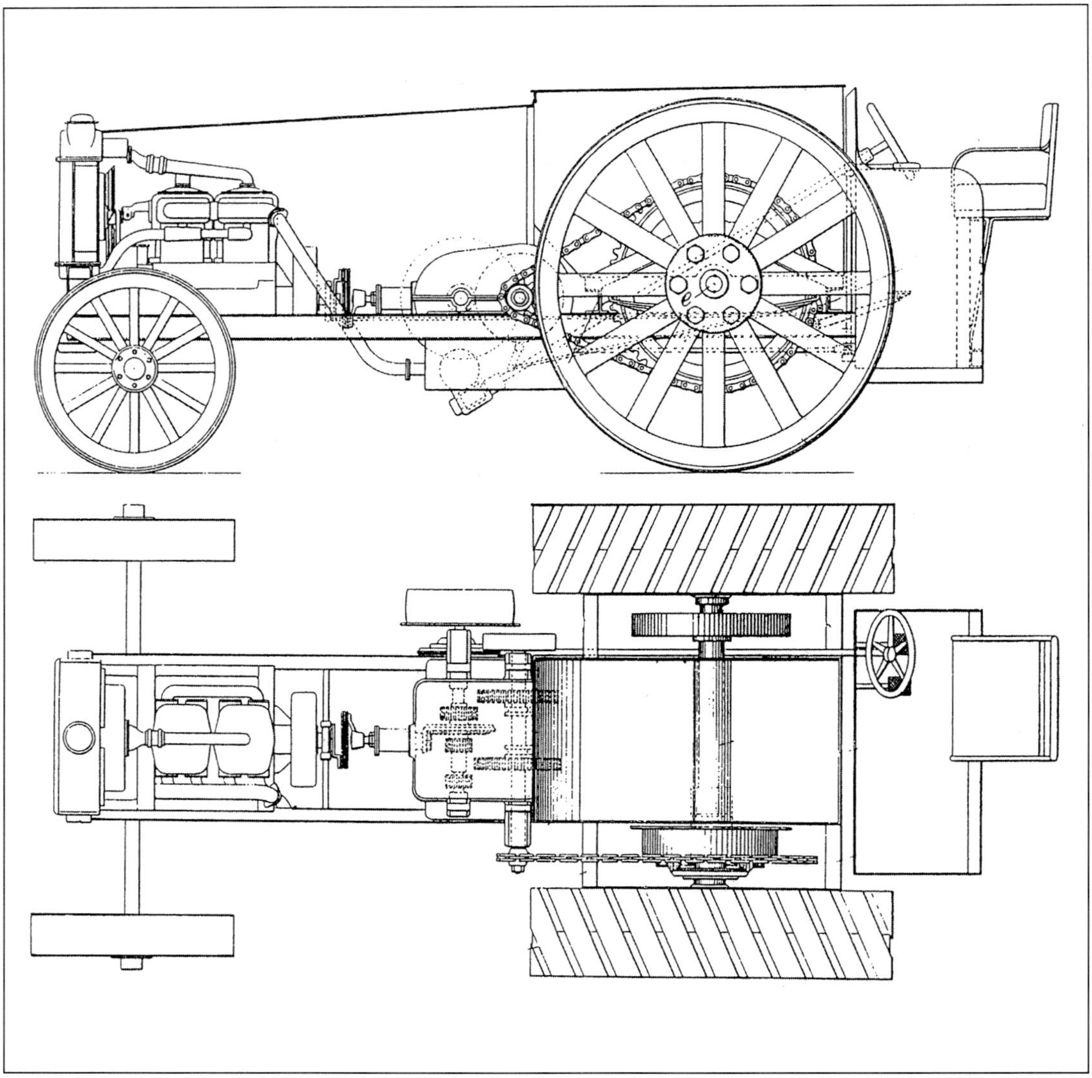

AGRICULTURAL TRACTORS

As the First World War continued to drag on more and more women were being employed as farm workers to replace young men being sent off to fight for King and country. It soon became apparent that the agricultural industry would be facing serious difficulties with this heavy drain of its manpower and the government began to actively encourage manufacturers to produce machinery to speed up farm work which up to this point had been carried out by time consuming horse drawn machinery. There was a flurry of petrol engined tractors introduced by various manufacturers and many trials were taking place to establish a machine which would get government approval much on the lines of the wartime subsidy type commercial vehicles.

In October 1917 Alldays introduced their agricultural tractor called the "General", the name implying the diverse uses to which this machine could be employed, not only for farm work, but as a heavy duty prime mover for road haulage. The design of the *All British* "General" was protected by six patents, number 114811 which covered the general design and layout, number 115277 which covered the drive mechanism, number 115398 dealing with the chassis construction, number 116192 covering the steering mechanism, number 122320 covering the design and construction of the road wheels and the final patent, number 123643, covering the design of tractor wheel treads.

The Alldays tractor had a wheelbase of 7ft 10ins and rear track of 6ft 3ins with 2ft 9ins diameter front wheels 6" wide and 5ft diameter rear wheels 12" wide. The power unit was a 25 h.p. four cylinder side valve engine fitted with a Zenith carburettor enabling the use of paraffin as a fuel which the tractor consumed at the rate of 2.25 gallons per acre when used for ploughing. The drive from the engine was taken through a leather cone clutch and a short drive shaft to a three speed gearbox located transversely across the chassis frame and then via chain to the rear axle. The three speed gearbox, of substantial proportions, gave speeds of 0.75 m.p.h. in first, 2.25 m.p.h. in second and 5 m.p.h in top gear. To the offside of the gearbox was a pulley enabling various pieces of static farm machinery to be driven with the tractor stationary and with ample power available to drive a 4'-6" threshing machine.

Immediately above the drive axle was a large tank which could be filled with 1,200 lbs of water to improve traction of the driven wheels, the tractor in this 'laden' condition weighed 55 cwts. The driver's seat was located well behind the rear axle and the worm and wheel steering gear actuated the front wheels via a roller and chain mechanism similar to that employed on traction engines. This combination of Ackerman type front axle coupled to traction engine type steering gave improved steering characteristics, particularly when ploughing and prevented the front wheels from twisting out of line and gave the tractor a turning circle of 26 feet. The front axle was sprung by a transverse multi-leaf spring pivoted at the centre to allow easy movement over rough ground.

One tractor operated by an engineering company, Messrs Palmer Brothers of Wisbech, was used as a road haulage machine which, by March 1919, had covered a distance of 5,000 miles and returned a fuel consumption of 6 m.p.g. running on paraffin. The "General" was designed to haul loads of up to 5 tons on ordinary roads at 5

ILLUSTRATION

This drawing shows the general layout of the "General" agricultural tractor. The forward mounted engine drove through a simple flexible coupling to the massive transversely mounted gearbox which then drove the rear axle by a chain. The engine was equipped with a gear driven water pump and belt driven fan, together with a large capacity radiator to ensure adequate cooling under all operating conditions.

The driving position to the rear of the tractor was quite low compared with some agricultural tractors and, although not shown here, a simple roof structure gave some protection to the driver from the elements. The braking system comprised a foot operated brake on the outside of the gearbox second motion shaft (just inboard of the driving pulley) and a screw operated brake on the rear axle which can clearly be seen inboard of the chain drive to the nearside rear wheel.

SEASONAL
PURPOSE

ALLDAYS
GENERAL PURPOSE
TRACTOR

LEFT-TOP: This photograph was published in 1918 and shows an Alldays "General" tractor towing trailers outside Birmingham Town Hall during a demonstration organised by the Women's War Agricultural Committee following the presentation of Certificates of Efficiency won by women in the Midland Land Workers Tests the previous Autumn. Women had become the main stay of the British agricultural industry following the loss of huge numbers of young men in the European killing fields of WW I and, in some respects, led to the more rapid mechanisation of the farming industry several years before it possibly would have happened in peacetime conditions.

(Photograph courtesy of Birmingham Library Services)

LEFT-BOTTOM: The 3 speed gearbox of the Alldays tractor showing the substantial proportions of the gears and casing itself. The plain bearings shown here were rapidly replaced by roller bearings, the gears and bearings being lubricated by oil contained in the lower half of the gearbox casing.

BELOW: A good view of the tractor complete with canopy above the driver's seat which was positioned much lower than in other contemporary tractor designs. "Country Life" said of this tractor, "In operation this tractor is capable of ploughing 3 or 4 furrows, and there is a striking absence of noise, vibration and fuss which one usually associates to a considerable degree with motor tractors."

m.p.h. Sales of these machines in the Birmingham area was undertaken by George Heath Ltd.

In June 1919 modifications were carried out to the design of the tractor which then became the MKI., the changes included omitting the water ballast tank and moving the driving position forward over the drive axle and the rear wheels could be specified as either 4ft 6ins diameter or 5ft diameter and could be specified with rubber stud tyres. Improvements were also made to the gear change mechanism. The wheelbase was altered to 7ft 6ins and the overall dimensions were 6ft 6ins wide and 12ft long and the tractor had a turning circle of 25 feet. The four cylinder engine, now fitted with an improved paraffin vaporiser, was rated at 25/30 h.p. and had a bore of 100mm diameter and stroke of 130mm. One of the new models was driven from Birmingham to the R.A.S.E. show at Cardiff in August 1919 hauling a trailer with a gross weight of 3 tons 7 cwts, completing the outward 100 mile journey in 19 hours and the return journey in 18 hours.

Development work went on rapidly and the MKII. version was also announced in June 1919 powered by a new overhead valve four cylinder engine with a bore of 4.3125 inches diameter and stroke of 6.5 inches which developed 30 h.p. at 1,000 r.p.m.

Prices for the new tractors, which were first exhibited at the 78th Royal Show in Cardiff, were £570 for the MKI, and £595 for the MKII. machine.

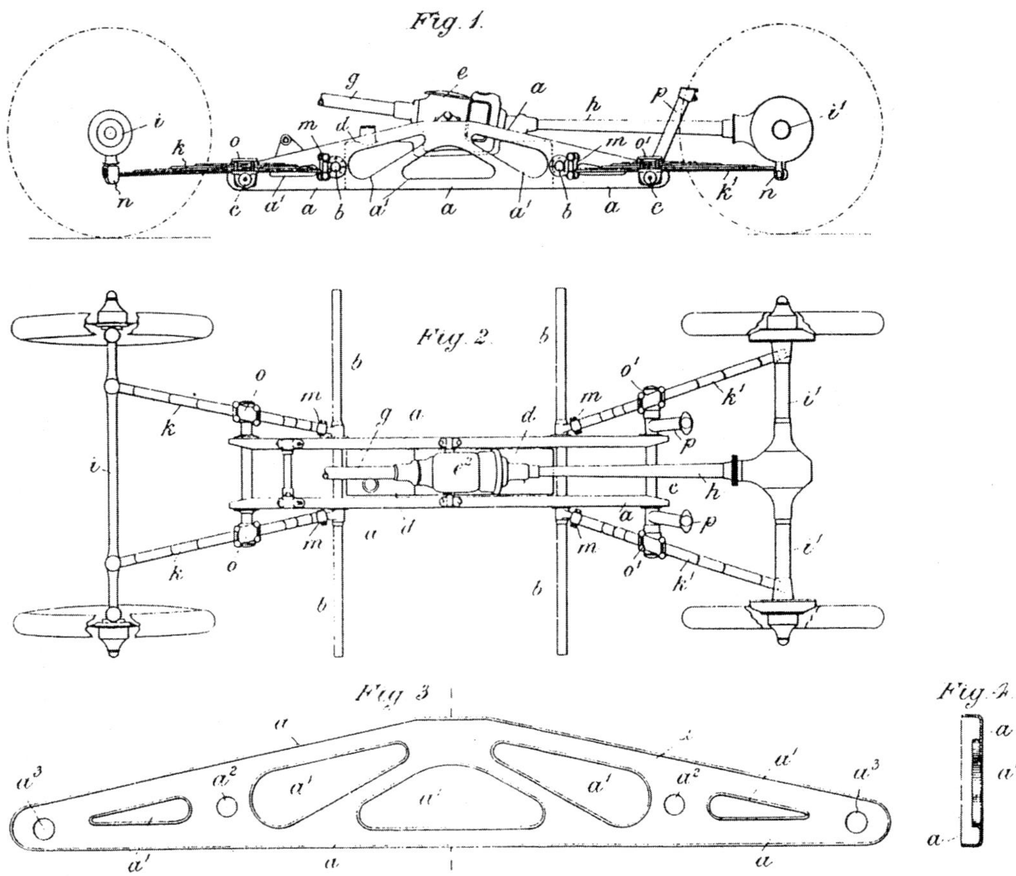

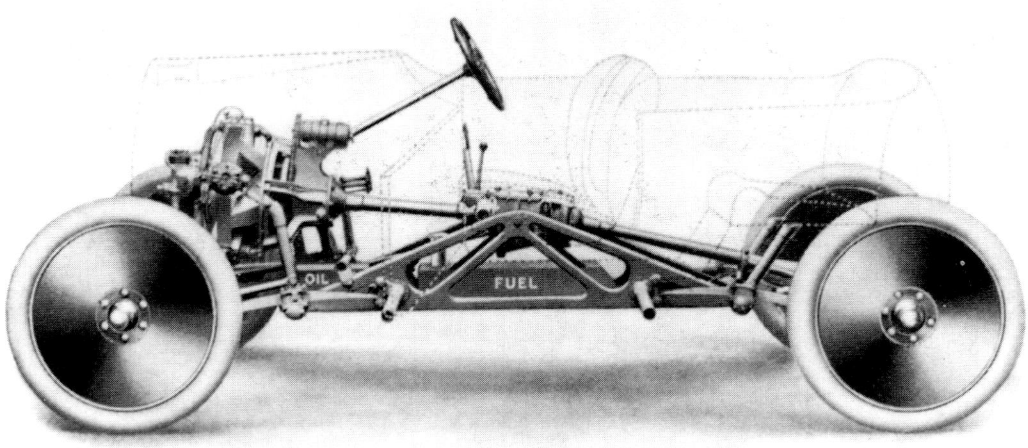

ELEVATION OF THE ENFIELD-ALLDAY CHASSIS SHOWING THE ELONGATED FRAME AND CANTILEVER SUSPENSION.
THE SHAPE OF THE FOUR-SEATED BODY IS INDICATED.

A REVOLUTIONARY CAR – THE BULLET

As the First World War ended the British motorist was looking for cheap motoring. Cheap cars to buy and cheap cars to run. Alldays were hoping to provide such a motor car in the Enfield - Allday 10 h.p. Light Car announced in November 1918.

This revolutionary car design was covered by no less than 8 patents and was designed by Arthur William Reeves, Manager of the Matchless Works. The engine was a 5 cylinder air cooled radial with aluminium cylinder heads, sleeve valves and aluminium/bronze pistons and was originally specified as having a bore of 62mm diameter and stroke of 80mm to give a capacity of 1,247 c.c. However, when the company actually produced a sales brochure the engine dimensions had changed to a 68mm diameter bore with 80mm stroke giving a capacity of 1,450 c.c. and the performance data had changed from 10.25 h.p. at 1,250 r.p.m. and 20.5 h.p. at 2,500 r.p.m. to 13.3 h.p. at 1,250 r.p.m. and 26.6 h.p. at 2,500 r.p.m. and the engine was reported to have run at up to 4,000 r.p.m.

The built up crankshaft ran in four roller bearings and carried the master con-rod which was connected to the number 1 piston, the remaining four con-rods being pivoted to the master, a practice quite common in aero engine design. The valve gear was comprised of two concentric sleeves operated by overhead rockers which opened and closed the inlet and exhaust ports in the detachable cylinder heads.

Engine cooling was achieved by a series of interconnected ducts which took in air via a centrifugal fan which replaced the conventional engine flywheel, and the light engine and chassis construction, together with the absence of a radiator and coolant helped to reduce the overall weight of the car to 9.75 cwts with a two seater body and 10 cwts with a four seater body.

The single dry plate clutch originally specified had been changed to a multi-plate clutch, with 13 plates, and the wheelbase of the car which had originally been stated as 8'-0" had been altered to 8'-9". The overall dimensions were 11'-9" long by 4'-8" wide. A three speed gearbox drove the car through a torque tube transmission to a spiral bevel rear axle.

Apart from the unusual choice of engine the car had a very unusual chassis frame made from two triangulated pressed steel side members between which was bolted the gearbox, together with two transverse tubular outriggers which passed through the sidemembers on which the bodywork was mounted. The petrol tank was mounted between the chassis sidemembers and below the gearbox with a forced feed to the carburettor.

The suspension comprised long multi-leaf springs which were almost flat and mounted asymmetrically on short transverse tubes which also passed through the sidemembers, the longer portion of the springs acting as cantilevers to the front and rear axles to give leading and trailing suspension respectively. To assist axle location two radius rods mounted parallel to, and above, the front road springs were used at the front and tie rods mounted on the rear axle tubes and connected to the torque tube located the rear axle and helped to eliminate side thrusts.

To accommodate different sizes of drivers the foot pedals and steering column could be adjusted to suit and the steering wheel itself

ILLUSTRATIONS

The top illustration is a drawing from patent number 157590 taken out by Arthur William Reeves in November 1919 to cover the design of the chassis frame for his 5 cylinder radial engined motor car. The basic construction consisted of two triangular shaped pressed steel panels as shown in Fig.3. through which two long body mounting tubes 'b' were welded, together with two shorter tubes 'c' at the outer extremities of the triangles to form the centre point mountings for the cantilevered road springs. The completed structure provided a fairly strong, but light chassis frame for a car which weighed only 10 cwts complete with 4 seater body.

The picture at the bottom of this page shows a completed motor car in chassis form based upon the Reeves design with the gearbox mounted in the top of the triangular chassis frame and the petrol tank mounted beneath it and with the propeller shafts between the engine, gearbox and rear axle clearly shown.

Front View of Enfield-Allday Chassis showing accessibility of all parts of the Engine. Magneto Carburettor, etc.

was designed to fold up against the column to provide extra room when getting in or out of the car.

The braking system included a novel cone shaped foot operated brake and more conventional internally expanding shoes, operated by rods and cables, in the rear axle drums.

A patented bulkhead panel formed the engine mounting and allowed a short drive shaft to be used between the engine and gearbox. This panel was also designed to allow the engine to be rotated about its crankshaft enabling each cylinder in turn to reach a convenient position in the engine compartment for servicing, the engine itself weighed about 123 lbs.

The car was expected to cost £275, but this had risen to £295 according to an advertisement published in "The Light Car & Cyclecar" dated February 15th 1919 and had risen again to £350 by the time the brochure was printed. In August 1919 the company had to announce that none of these revolutionary new cars had been delivered to customers, who were queueing up to buy this car which promised exceptionally low running costs with respect to fuel consumption and tyre wear. The car, according to the company, was still undergoing rigorous development testing in Scotland.

One year later, in August 1920, it was announced that production of "The Bullet" had been abandoned due to the increased production costs which had sent the price of the car soaring to £550, double its originally quoted price!

However, there is a widely held belief that the car had been abandoned, not just because of cost, but because of persistent and incurable reliability problems with the engine and Alldays could not risk their good reputation by allowing these cars to get into the hands of customers, many of whom had already had many years' experience of good quality products from the company.

In reality the engine had been designed at a time when everything was against its success. Air cooled engines were far from reliable anyway, but to add the radial configuration and aluminium cylinder heads with sleeve valves into the equation was courting disaster as metal and lubrication technology of the day would not have supported such an advanced engine design and it was virtually doomed from the day the designer's pen drew it up on paper.

It is a fact of life, that various motor car designs appeared in the early days of motoring which failed at the time simply because they were too advanced for the technology then available, but when resurrected years later would be hailed as something 'new'!

In September 1920 A.W.Reeves, the car's designer, resigned from the company and announced his intention to set up his own company to produce another air cooled engined car made to his patents, to be known as the "Reeves Radial".

Alas, nothing more appears to have been heard of Mr Reeves or his ingenious motor car designs, but "The Bullet" had cost Alldays a lot of money in development costs and tooling up for its production and none of this could be recouped from sales. Indeed, the company had been forced to return money from customers who had paid deposits on cars in advance which had probably done little for customer relations at a time when all car makers were having to fight hard to keep their share of a very competitive post war market and, to make matters worse, they had 'put all their eggs into one basket' and had nothing left to put into production!

PHOTOGRAPHS

The top picture shows a front view of "The Bullet" motor car in chassis form with an excellent view of the 5 cylinder radial, air cooled engine. Four patents were taken out by Reeves to cover the design of this engine, numbers 137439, 154344, 154423 and 158371, to cover the ducted cooling, cylinder head cooling, cylinder design and the sleeve type valve gear respectively.

This picture also shows the front road springs which were of leading cantilever design together with radius rods above the springs to assist with axle location and to the rear can be seen the rear axle with the brake drums clearly visible, together with the two transverse tubes used for body mounting.

The lower picture shows a completed 4 seater motor car which had quite a pleasant conventional appearance hiding its revolutionary design. Had the motor car been a success Alldays may well have become a high volume motor car manufacturer well into the 1930s, but it was a total disaster which led directly to the financial collapse of the company.

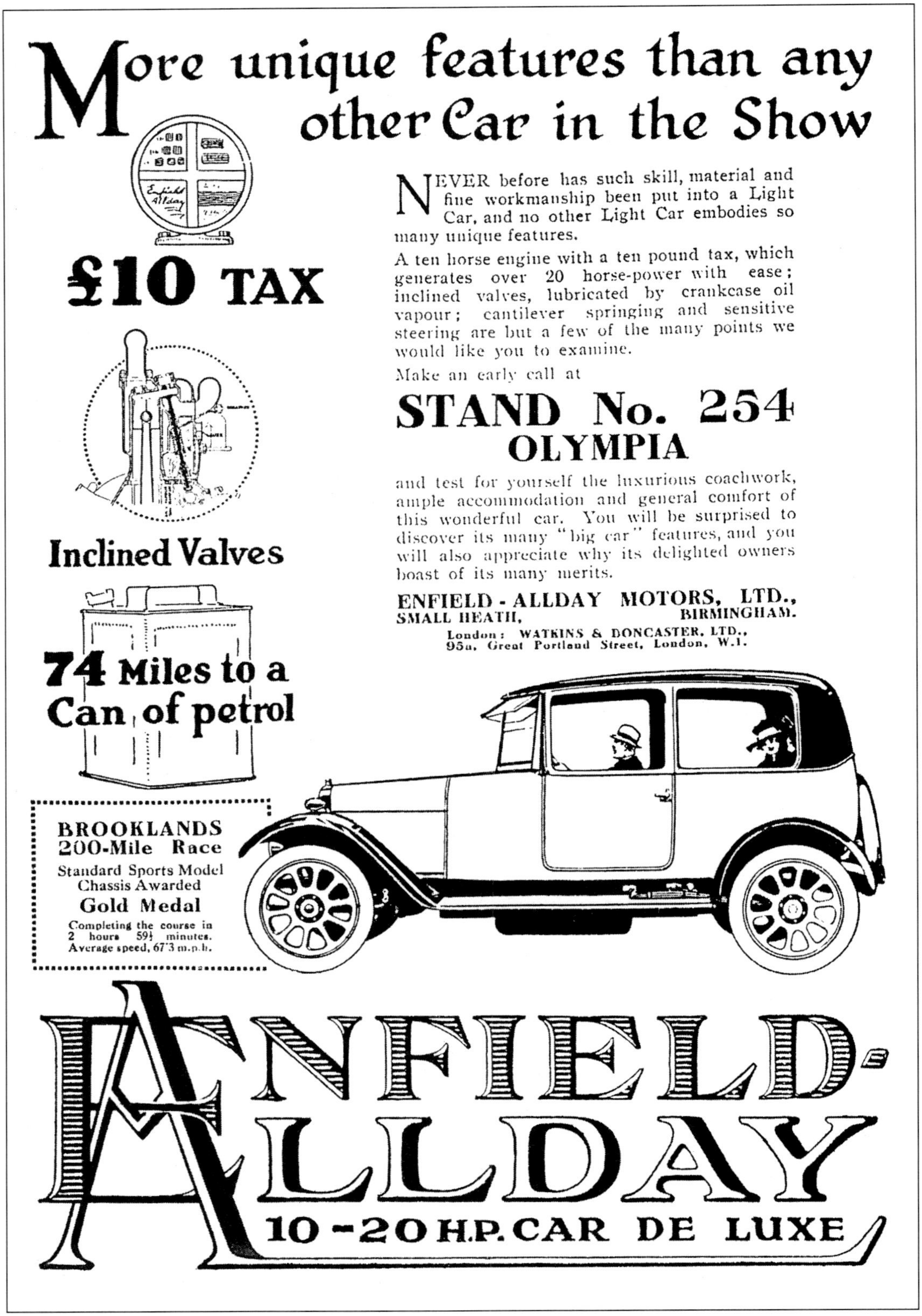
More unique features than any other Car in the Show
£10 TAX
Inclined Valves
74 Miles to a Can of petrol
NEVER before has such skill, material and fine workmanship been put into a Light Car, and no other Light Car embodies so many unique features.
A ten horse engine with a ten pound tax, which generates over 20 horse-power with ease; inclined valves, lubricated by crankcase oil vapour; cantilever springing and sensitive steering are but a few of the many points we would like you to examine.
Make an early call at
STAND No. 254
OLYMPIA
and test for yourself the luxurious coachwork, ample accommodation and general comfort of this wonderful car. You will be surprised to discover its many "big car" features, and you will also appreciate why its delighted owners boast of its many merits.
ENFIELD - ALLDAY MOTORS, LTD.,
SMALL HEATH, BIRMINGHAM.
London: WATKINS & DONCASTER, LTD., 95a, Great Portland Street, London, W.1.
BROOKLANDS
200-Mile Race
Standard Sports Model Chassis Awarded
Gold Medal
Completing the course in 2 hours 59½ minutes. Average speed, 67·3 m.p.h.
ENFIELD-ALLDAY
10-20 H.P. CAR DE LUXE

RETRENCHMENT IN 1923

Prior to the First World War plans had been set in motion to jig up the 10/12 h.p. motor car for volume production, but the war had shelved these plans and, in the intervening period, experimental work had begun on the air cooled radial engined car designed by A.W.Reeves. Despite reliability problems with this new car, in 1921 the Alldays members of the company's management had been outvoted to volume produce the pre war car in favour of Reeves car. As a direct result of this decision all the jigs and fixtures for the pre-war car were sold off to William Morris in Oxford and one can only speculate as to what part this played in helping Morris to set up his motor empire. However, this sadly led to the resignation of Percy Allday and, shortly afterwards, Percy's father Edmund decided to retire, so ending all connections of the Alldays family with the company which still bore its respected name.

With production of Reeves' car abandoned the Allday policy to produce the pre-war car would appear to have been the correct route to have taken, but the two key members of the management had gone and the company faltered on. About this time two Italian brothers, the Bertellis, were employed by the company, probably as consultants, to design and put into production a more conventional car than the ill fated Reeves Radial. The mechanical parts of the new car, the LC11 type 10/20 h.p., were designed by A.C. Bertelli, more popularly known as "Gus", and the bodywork designed by Harry Bertelli. Production commenced in 1921 starting at chassis number 10,001 with, unusually for Alldays, the engine number matching the chassis number. The first four cars were built specifically for the Olympia Motor Show and the company records show the first 'production car' as being chassis number 10,005. The chassis was of pressed steel channel construction which was swept upwards over the rear axle with the rear suspension being of the trailing cantilever type. Goodyear steel detachable wheels fitted with Dunlop 'Magnum' tyres were standard across the range, as were Serck radiators, Zenith carburettors and Brolt lamps, dynamo and starter. Magnetos appear to have been a mixture of T.B., Lucas, Watford and M.L., with the occasional B.T.H. or Simms thrown in! A.C. Bertelli took delivery of one of the new cars himself in June 1921, chassis number 10,011 fitted with a special 3 seater sports body with Peacock Blue paintwork and red trim, louvred bonnet and external exhaust pipes, and registered OH 8712. (This is believed to have competed at Shelsley Walsh.) About 83 cars were built in 1921, but customers soon discovered problems with rear axle crown wheel and pinions failing, clutch and gearbox problems and even chassis frames cracking.

Despite production and financial problems a works 10/20 h.p. Standard Sports Model was entered in the 1922 Brooklands 200 Mile Race completing the course to come in 4th position with an average speed of 76.88 m.p.h. and awarded a Gold Medal. This car was powered by a 4 cylinder engine with a bore of 63.5mm diameter, stroke of 117.5mm, 4 valves per cylinder, and developed 40 b.h.p. at 4,000 r.p.m. A similar car was entered in the Isle of Man 1,500 c.c. race and was the only all British car to finish the race.

By February 1922 the matching of chassis and engine numbers was abandoned and the company records became poorly maintained,

ILLUSTRATION

This advertisement for the 10/20 h.p. motor car appeared in "The Motor" magazine dated November 2nd 1921.

The petrol consumption of "74 miles to a Can of petrol" works out at 37 miles per gallon, the petrol can in those days containing 2 gallons. The motor car illustrated is pleasantly styled and, in fact, Alldays cars at this time won many prizes in Concours competitions held all over the country. The diagram showing a section of the engine illustrates the fixed cylinder head and inclined side valves currently being used on the greatly reduced Alldays range of cars.

By this time car production was being undertaken mainly from the Waverley Works which had become the home of another Alldays company, Alldays Motors Repairs Ltd., which, as the name suggests, was the works repair centre for company produced motor vehicles.

PHOTOGRAPHS

TOP: A group of three Friction Drop Stamps driven by an electric motor mounted to the left of the supporting steel framework. The drop stamps were, from left to right, a 15 cwts model which was 19" wide between the vertical guides, weighed 15 tons and cost £414; a 7 cwts model which was 14.5" between the guides, weighed 8.5 tons and cost £235 and a 10 cwts model which was 16" between the guides, weighed 10.75 tons and cost £358. All three drop stamps had a stroke of 7 feet and the range also included models of 3 cwts, 5 cwts and 20 cwts, which cost £144, £184 and £534 respectively.

LEFT-BOTTOM & RIGHT: A three rotor Water and Air Cooling Machine with its top casing removed to show the three rotors.

A diagram showing one of these machines coupled to a Serlano fan is shown to the right.

These machines were designed primarily to cool the coolant from gas or oil engines without having to resort to the use of cooling tanks requiring a large space outside an engine room together with all the necessary plumbing. Each rotor comprised a central disc mounted on the drive shaft with radial arms supporting a spiral made from a continuous galvanised steel plate. Water from the gas or oil engine entered into the base of the cooler which was made from two cast iron ends supporting a mild steel plate tank and the main drive shaft for the rotors. A thin film of water was picked up by the rotors as they were slowly rotated through the water in the lower half of the casing and air was blown through the spirals of plates causing almost instantaneous heat loss with very little loss of the actual coolant water. These machines were made in fourteen different sizes, the smallest having rotors 22" in diameter and measuring 25" wide by 33" long by 24.5" high and the largest machine had rotors 8' 10" in diameter and measured 9' 3" wide by 6' 1" long by 9' high.

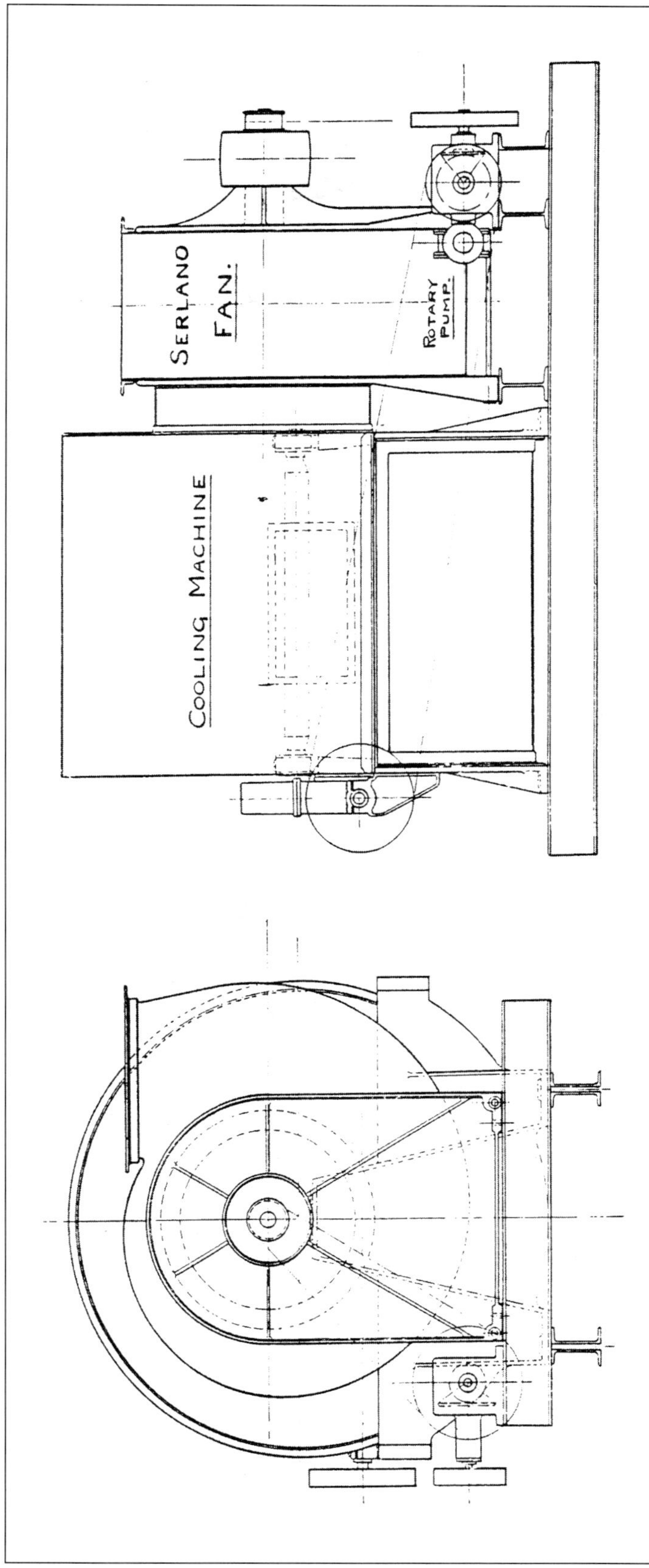

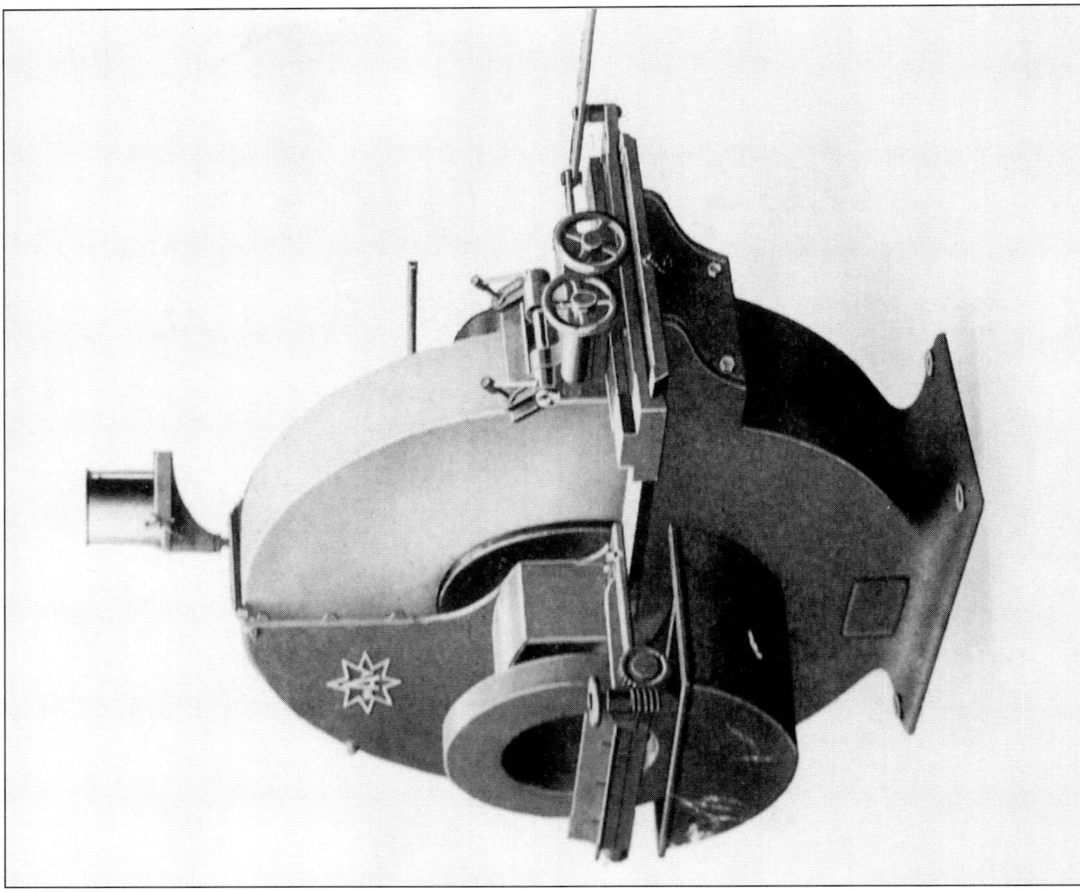

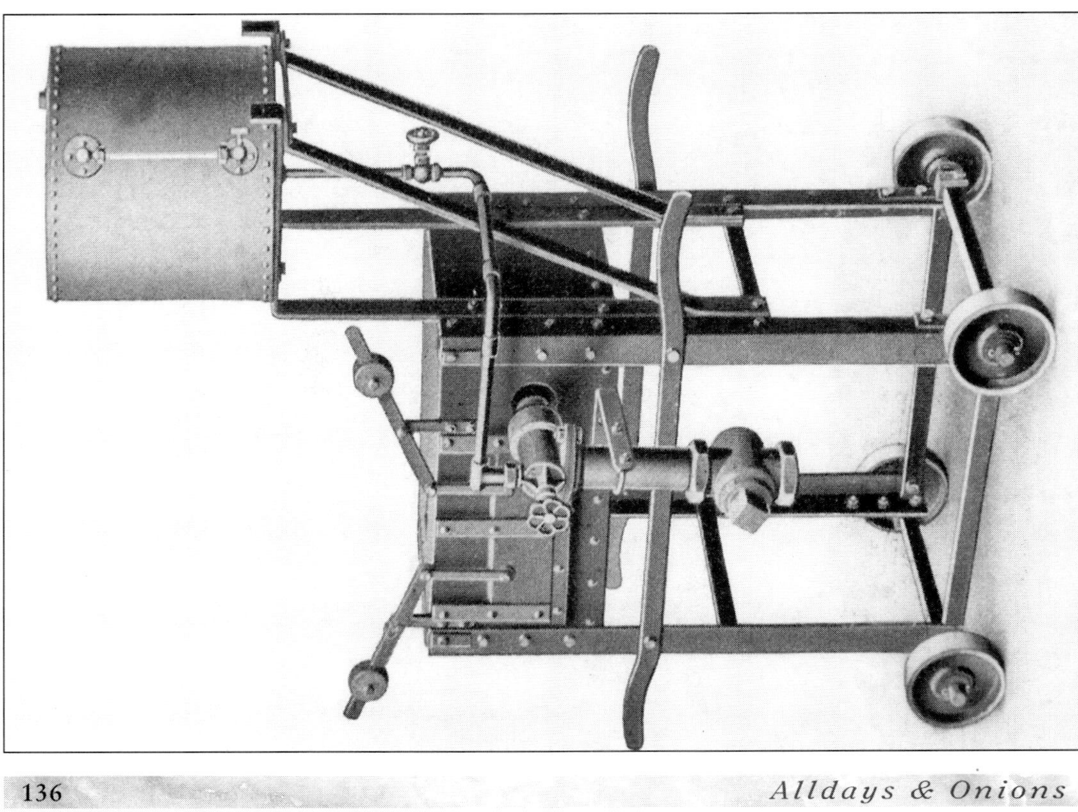

LEFT-TOP: A "Paragon" Grindstone complete with compound slide rest across the front of the main grindstone which was 60" in diameter by 9" wide. Mounted on the left of the main grindstone was a "Water-of-Ayr" stone 24" in diameter by 4" wide used for putting a fine edge on knives,etc. Belt driven at 62 r.p.m. and weighing 2 tons 3 cwts it cost £85-10s complete.

LEFT-BOTTOM: One example from a range of six "Empire" oil fired Rivet Heating Furnaces, but this particular furnace was also portable. Made from steel plate and lined with fire bricks, the rivets were fed into the furnace through a hole in the top and could heat about 300 .75" diameter rivets per hour and consume about 2.25 gallons of oil. Although portable, these furnaces still required an air blast from a blower. The furnaces cost £18 each.

BELOW: This unusual looking railway truck was patented in 1914 by John Newton Collins, Manager of the Fan Dept., and Frank Lyon, a Draughtsman, both of the Great Western Works, and was designed to distribute stone dust in mine tunnels to render the deposits of coal dust non-inflammable. After filling the trucks hopper with suitable stone dust it would be attached to an underground colliery train where the dust would be distributed automatically by the fan which was gear driven from the axle of the truck. Alldays even supplied a 4 ton sample of inert stone dust before the potential customer bought a truck!

indicating, perhaps, the general decline in the company. The motor car range in 1922 comprised basically two models, a 10/20 h.p. and 12 h.p., but with a variety of body styles, which included the 10/20 h.p. in 2 seater, 4 seater and Sports model at £445, £450 and £475 respectively. The 12 h.p. model was available as a 2 seater, 4-5 seater, Folding Coupé, All Weather 4-5 seater, Limousine Coupé, Landaulette and Saloon at prices of £445, £450, £575, £595, £650, £650 and £695 respectively. For 1923 the 10/20 h.p. model was retained and an improved 12 h.p. model, now designated as the 12/30 h.p., introduced with a 4 cylinder engine with a bore of 69mm diameter and stroke of 117.5mm with a capacity of 1757 c.c., both engines featuring inclined side valves and non detachable cylinder heads. The 10 h.p. car was based on a 9'-0" wheelbase chassis with a 3 speed gearbox and spiral bevel rear axle and the 12 h.p. model was based on 9'-6" wheelbase chassis with a 4 speed gearbox.

Changes in the management of Enfield-Alldays Motors Ltd. had little affect on the financial position and the name of the company was changed again to Enfield-Alldays Motors (1923) Ltd. In the following year the parent company also ran into financial difficulties chiefly, it was thought, by the £120,000 spent on racing and "The Bullet" car.

A Receiver was appointed to run the company, Mr C.Herbert Smith, resulting in the closure of "Matchless" and "Waverley" works and the transfer of all motor car, commercial vehicle, motor cycle and cycle work to the Great Western Works.

Percy Allday, apparently, tried in vain to buy back his old company from the Receiver and shortly afterwards started his own company, William Allday & Co.Ltd., with offices in Queens College buildings, Paradise Street, Birmingham and works in Blackheath, making general smiths' and foundry equipment.

Contractors to the Admiralty, War Office, Post Office, India Office, Etc.

Telegrams: "Alldays, Birmingham"
Telephone No.: 487 Victoria (P.H.)

Codes:
A.B.C. 4th & 5th Editions
Marconi International

London Office: 40/44 Holborn Viaduct, E.C.
Telephone No.: 2032-3 Holborn
Telegrams & Cables: "Typhoon, Cent, London"

RECEIVER AND MANAGER OF

ALLDAYS & ONIONS LIMITED

Specialists in Smithy and Foundry Plant

GREAT WESTERN WORKS
SMALL HEATH
BIRMINGHAM

All communications to be addressed to the Company

REFERENCE HEW/FP.

ESTIMATE.

13. 2. 25.

Dear Sir,

With reference to your enquiry over the telephone to-day, we have pleasure in quoting as follows:-

COMBINATION BRAZING FORGES, without blowpipes or rubber tubing:-

Patt. No.	Price. Each. £.	s.	d.	
C1.	4.	5.	0.	)
C2.	4.	17.	6.	)
C3.	5.	10.	0.	) LESS 5%.
C6.	6.	12.	6.	)
C7.	8.	0.	0.	)

Extra for top plate if required - 5/- per forge less 5%.

TERMS:—2½% MONTHLY ACCOUNT
Delivered f.o.r. Manchester.
Time - we can give early delivery.

Yours faithfully,

P. P. ALLDAYS & ONIONS LIMITED,
C. HERBERT SMITH,
RECEIVER AND MANAGER.

NEW ALLDAYS & ONIONS, LTD.
ESTABLISHED 1650

Telegrams: "ALLDAYS, BIRMINGHAM."
Telephone: 487 VICTORIA.

LONDON OFFICE: 40-44 HOLBORN VIADUCT, E.C. 1.
Telephone: 2032 HOLBORN.
Telegrams & Cables: "TYPHOON, CENT, LONDON."

GREAT WESTERN WORKS,
BIRMINGHAM.

16th March, 1926.

References
Yours WVH/AS.
Ours JL/AB.

Directors:
Sir W. D. Mitchell Cotts, Bart., K.B.E.
A. T. Cooking, J.P.

ESTIMATE.

Dear Sirs,

We thank you for your enquiry of the 12th inst., and have pleasure in quoting as under:—

1 G.2. "Climax" Roots Patent Rotary Hand Blower with 1¾" diameter outlet, chain driven.

Price.............................. £9. 0. 0 nett.

Terms:- 2½ Monthly A/c
f.o.r. Birmingham, packing extra.

Time:- 3 to 4 weeks.

Yours faithfully,

P.P. NEW ALLDAYS & ONIONS LIMITED.
Wilfred H Goodwin
SECRETARY.

Messrs. William Allday & Co. Ltd.,
Pneumatic Engineers,
38a, Paradise Street,
Birmingham.

NEW ALLDAYS & ONIONS LTD.

On July 8th 1925 "New Alldays & Onions Ltd". was formed under the ownership of Sir William Mitchell Cotts and Mr. A.T. Cocking J P. Sir William Mitchell Cotts and Partner had purchased several other companies and many of their products were to be manufactured in the Sydenham Road factory to replace the rapidly declining motor car, commercial vehicle and motor cycle part of the company's business. These companies included Britannia of Coventry Foundry Company (moulding machines and foundry equipment), Neals of Kettering (sand mills and mixers), Peter Pilkington of Bamber Bridge, Lancashire (pneumatic hammers and compressors), Samuelsons of Banbury (Roots type blowers) and Thwaites Brothers Ltd. (cupolas, ladles, steel converters, Roots blowers and steam hammers).

These acquisitions caused their own problems at Sydenham Road where vast quantities of drawings and patterns from these companies needed to be stored. The "Great Western Hall" built by the company on the opposite side of the road in 1896 as a Mess Room had been closed for some time and was now to be given a new lease of life as a Pattern Store.

The Thwaites steam hammers covered a range of 2½ cwts up to 15 tons and included double standard tilting hammers, suspension type hammers, open frame hammers and single frame hammers. The largest hammer available, the 15 ton suspension type had a cylinder 42" in diameter, a stroke of 96" was 18 feet wide between the pillars, had a hammer which weighed 82 tons and an anvil block which weighed 120 tons. Smaller Thwaites "Rigby" type steam hammers were also in production in a range of 1½ cwts up to 2 tons.

Pneumatic hammers driven by electric motors were made in 8 models from ½ cwt to 10 cwts with prices from £133 to £455. Spring type double Oliver hammers and pneumatic planishing and riveting hammers still remained in the product range which also included hand or foot operated automatic hammers and friction lift drop hammers.

Fans of all types and sizes were still in the catalogue of the new company and these included the "Serlano" type in sizes from 960 cu.ft/min to 22,500 cu.ft/min; "Waverley" as blowers from 398 cu.ft/min to 10,250 cu.ft/min, exhausters from 164 cu.ft/min to 4,700 cu.ft/min and dust extractors from 200 cu.ft/min to 5,700 cu.ft/min. The "Serlano" fans were also available in steel plate designs for ventilation from 830 cu.ft/min at £15-17s-6d up to 19,000 cu.ft/min at £51. The "Sertus" range of fans mounted in a steel frame was available in sizes from 12" diameter to 60" diameter covering a capacity of 1,900 cu.ft/min to 50,000 cu.ft/min and were made in belt driven form or directly driven by electric motors.

A completely new range of blowers and exhausters was introduced of Thwaites design which included belt driven low pressure positive blowers in sizes from 25 cu.ft/min to 25,000 cu.ft/min., 'HS' type high speed low pressure blowers and exhausters directly driven by electric motors from 11 cu.ft/min at £18-17s-6d to 9,940 cu.ft/min at £277 and 'B' type belt driven high pressure positive blowers from 1,200 cu.ft/min up to 25,000 cu.ft/min. Alldays "Empire" rotary blowers and exhausters and "Duplex" blowers for cupolas and smiths' hearths were also still in production together with "Mota" electrically driven forge fans.

LETTERS

LEFT: The two letters reproduced here, whilst of little technical significance, are historically quite interesting. The first one with the company name in large print has been overstamped "Receiver and Manager of", having been written when the company was in the hands of the Receiver, Mr C. Herbert Smith.

The second letter was written in the next stage of the company's history with the name overstamped "New" Alldays & Onions Ltd. However, the real interest in this letter is that it was dealing with a request for the supply of a "Climax" rotary hand blower to the company William Allday & Co.Ltd. It appears somewhat ironic that Percy Allday having severed his links with the old family business and then starting up his own business had to order items from his old company, which was a competitor in the same field of general engineering!

Illustration from photograph of Lip Pouring Foundry Ladle, with lifting link astride of beam.

Despite the introduction of belt and electrically driven blowers a range of bellows was still being made which included long shape smiths, "Reversible Pipe" smiths, double blast circular, horseshoe double blast, moulders, electricians and the humble house bellows still being made in sizes from 8" at £1-16s-0d per pair up to 18" at £10 per pair.

Alldays designed Smiths hearths were also still being made in the "Climax" range from 30" square at £9-4s-9d up to 78" diameter annular hearths at £60, and "Caledonian" models in sizes from 36" square at £6-13s-9d up to 48"x72" at £20. Portable fan forges had always been popular products made by the company and a whole range was still available from the new company with brand names such as, "Monarch", "Ajax", "Sydenham", "Ariel", "Era", "Century", "Typhoon", "20th Century" and the long established "Branston" model. Seventeen different models of rivet forges were in production both hand and treadle operated, together with 9 brazing forges.

The "Empire" range of products now included gas and oil fired hardening and annealing furnaces and forging furnaces. Oil fired steel tyre heating furnaces were in production capable of handling locomotive wheels from 20" diameter up to 72" diameter at about £100 and oil fired tilting melting furnaces and rivet heating furnaces were also available, the latter in a range of six models from £21 to £50-10s-0d. "Empire" and "Climax" oil burners were also being made suitable for industrial boilers and furnaces, together with a patented gas burner for similar applications.

Thwaites designed cupolas had now replaced the Alldays designs, their "Rapid" range in 3 basic types covered 1 ton up to 15 tons and was augmented by two "Dwarf" cupolas 27" diameter and 36" diameter intended for smaller factories. The 1 ton "Rapid" cupola was 2ft 6ins in diameter and 22ft 4ins high and weighed 2 tons and the largest 15 ton model was 8ft in diameter, stood 40 ft high and weighed 16½ tons. Hand operated and electrically operated charging machines were now available for feeding cupolas both large and small, together with all the necessary staging and lifting gear to make a complete installation. Geared crane ladles in sizes from 10 cwts at £18 up to 5 tons at £74-18s-0d were in the standard range and special sizes could be made to order.

The company's range of cycles had now been reduced to four basic models, a tradesman's box tricycle, a tradesman's cycle, a military/GPO cycle and a standard bicycle.

The "Allon" 2 stroke motor cycles had been abandoned and a completely new "Alldays" 350 model introduced powered by a 349 c.c. single cylinder oil cooled Dorman-Bradshaw engine with overhead valves.

Sand and mortar mills became an integral part of the new company's product line up, the latter with rolls up to 48" diameter by 18" wide running in a 9ft diameter pan weighed over 11 tons and cost £234. "New Century" stone breakers with capacities of 32 tons to 80 tons per day were also being made and cost £48 to £134.

Although many engineering companies had been reduced to working as little as 3 days in a fortnight by the economic conditions then prevalent, Alldays were managing to work 5 days per week from 9.00 a.m. to 4.00 p.m. and the still surviving Alldays Motor Repairs Ltd. was working a 57 hour week under the managership of Mr Alan Cocking who was the son of Mr A.T.Cocking. (This part of the company eventually closed down about 1933/34 when the demand for spare parts diminished chiefly as a result of laws being introduced banning the use of commercial vehicles with solid tyres and financing

PHOTOGRAPHS

Both of these pictures help to illustrate a significant change in the company for, as well as the name change to New Alldays & Onions Ltd., the product range was being updated, not with Alldays designs, but with products 'acquired' by taking over other companies. The top picture shows a Thwaites designed geared foundry ladle and the lower picture a Thwaites designed steel convertor, but of a type previously not available from the Alldays company who had concentrated on cupolas.

ALLDAY'S
CARRIER TRICYCLE

ALLDAY'S
CARRIER
CYCLE

PHOTOGRAPHS

LEFT: Examples of the Carrier Tricycle and Carrier Cycle still being made by the company in 1930. The tricycle featured a frame built in accordance with a design patented, number 3454, as early as 1909. By comparing this photograph with the picture of a carrier tricycle of 1900 shown earlier in this history, it will be seen that the patented frame is of more substantial construction and also featured an improved front axle and steering mechanism allowing the carriage of much heavier loads. The carrier cycle was extremely popular with local shopkeepers who, even into the 1950s, were quite prepared to deliver orders to their customers free of charge!

BELOW: One of the last examples of GPO cycles made in 1930. With various additions this cycle also formed the basis for the military cycle supplied in large numbers to the British Army.

a development programme to convert their once huge range of vehicles to pneumatic tyres was considered prohibitive.)

Hoists, winches and cranes had also been part of the company's standard range of products and the new company continued to expand this range and still produced a full range of tools for engineering and smiths' work.

One range of products rarely associated with the company was the manufacture of pyrometers and thermocouples to measure and record the temperature inside furnaces. Wall mounted pyrometers complete with 36 feet of cable and a 3 feet long thermocouple were made in ten different temperature ranges from 150°C to 2,500°C, the cheapest model covering a range from 150°C to 600°C cost £9-3s-3d and the top of the range model covering 800°C to 2,500°C cost £12-7s-0d.

Portable pyrometers were also made with the indicator mounted in a polished wood case. Made in a similar range to the wall mounted models, these cost from £10-13s-6d up to £13-13s-0d. A portable Fixed Focus pyrometer which allowed the temperature to be checked without inserting a thermocouple into a furnace and covering a range from 700°C to 1,400°C cost £24-13s-6d.

To record furnace temperatures a wall mounted 10" diameter 8 day clockwork recorder was made with replaceable 24 hour charts. The recorder cost £28 and the charts cost 17s-0d per 100.

In 1927 some parts were made for 3 cwt ships' davits and in 1936 even humble Bunsen Burners were produced! Items for metal/heat treatment such as Tinning Baths, Salt Baths, Cyanide Baths and Oil Tempering Tanks were produced and machinery such as Rivetters, Stonebreakers, Sand Mills, Mortar Mills and Concrete Mixers introduced, together with a new range of Grindstones from 21" up to 4 feet. Fans were produced in bewildering types and numbers which included the Serlano range in sizes of $7\frac{1}{2}$", 9", 12", 15", 18", 21", 24", 27", 30", 36", 42", 54" and 60" and, in 1925, some parts were also made for Vacuum Cleaners.

LIMITED.
VULCAN IRON
WORKS.

PHOTOGRAPHS

LEFT-TOP: The formation of the new company saw the abandonment of the manufacture of all internal combustion engines previously made at the works, including the world famous "Allon" two stroke motor cycle engine. However, a completely new motor cycle was soon made available known as the Alldays 350 and featuring a 349 c.c. overhead valve single cylinder Dorman Bradshaw engine which was oil cooled. The engine had a bore of 68mm diameter and stroke of 96mm with an aluminium piston, detachable cylinder head, B.T.H. magneto and Binks or Amac carburettor. Fitted with a three speed Sturmey-Archer gearbox and chain drive the new machine was said to be capable of up to 70 m.p.h. and cost £55.

LEFT-BOTTOM: The lower photograph shows a Thwaites designed steam hammer for the fitting and removal of steel tyres on railway wheels, which was a valuable piece of equipment in many railway engineering workshops around the world.

ILLUSTRATION

RIGHT: The 'New Century' Stone Breaker was made in a range of six sizes capable of handling 32 cwts to 80 cwts per day. Some idea of the size of these machines can be gathered from some dimensions and weights quoted in the catalogue. The smallest machine had a pulley 19.5 ins in diameter by 4 ins wide and the whole machine weighed 1 ton 10 cwts and the largest machine had a pulley 30 ins in diameter by 8 ins wide and weighed 4 tons. The stone breakers cost between £48 and £134 and could also be specified mounted on a wheeled cart suitable for being drawn by a horse for an extra £22.

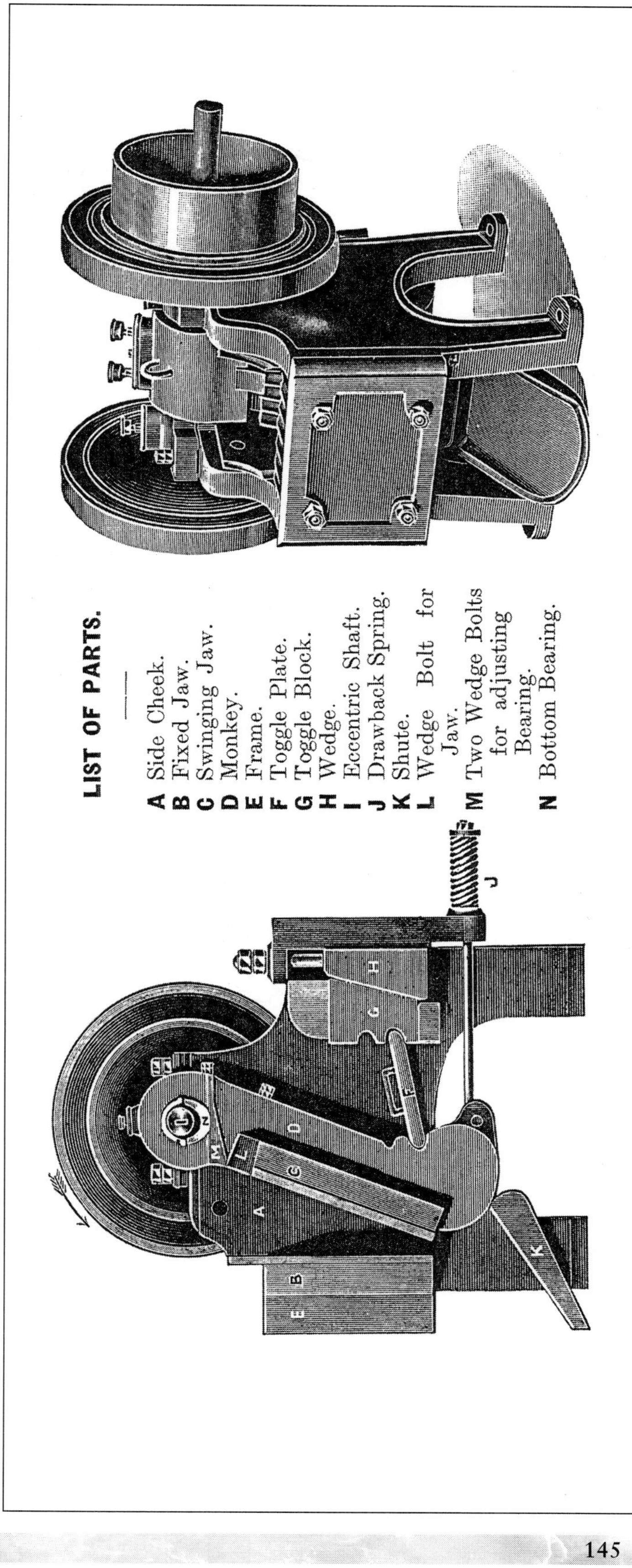

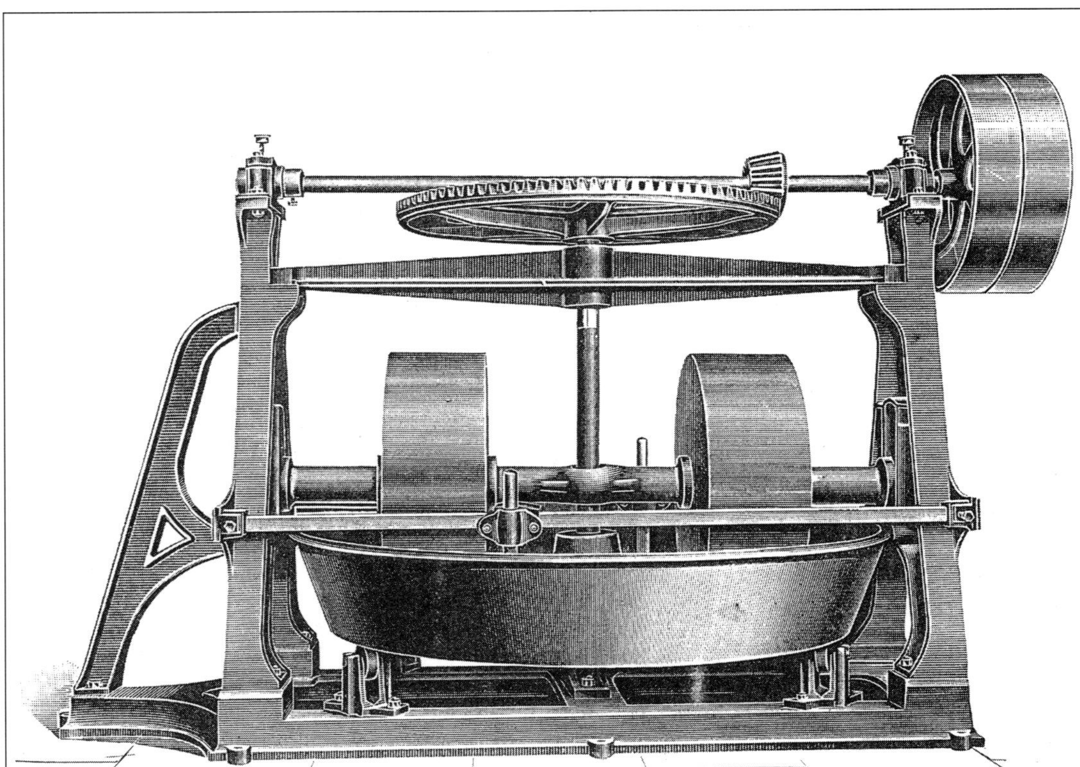

ILLUSTRATIONS

LEFT: A large range of Mortar Mills was made including this portable model mounted on a wheeled truck shown at the top of the page.

The smallest model had a pan 5 feet in diameter with the two rolls 30 ins in diameter by 10 ins wide and with a drive pulley 24 ins in diameter by 5 ins wide and weighed 3 tons 5 cwts and cost £78. The largest portable model had a pan 9 feet in diameter with rolls 48 ins diameter by 18 ins wide and with a drive pulley 42 ins in diameter by 9 ins wide, weighed 11 tons 15 cwts and cost £246.

The lower illustration shows an over-geared Mortar Mill made with pans from 4 feet in diameter up to 9 feet in diameter and with similar sized rolls and pulleys to the portable model. These static machines cost £46 to £234.

RIGHT: This shows a hand operated screwing machine for threading pipes and tubing. This machine was available in sizes which could cut threads on pipes up to 2 ins in diameter and cost £13 as a bench model and the largest machine which could cut threads on pipes up to 4 ins in diameter and cost £28, plus £4-5s for the cast iron stand. These machines could also be supplied with a 3 speed cone or fast and loose pulley to enable them to be driven by a belt from an overhead lay-shaft for an extra £2-8s.

The "Octopus" Chain Pipe Wrench was another tool greatly valued by pipefitters and was made in eight sizes from the smallest for pipes up to 0.75 ins diameter, which was 13.75 ins long and cost 10s-6d each, up to the largest model for pipes up to 18 ins in diameter, which was 7ft 3ins long, weighed 137 lbs and cost £8-6s-8d each.

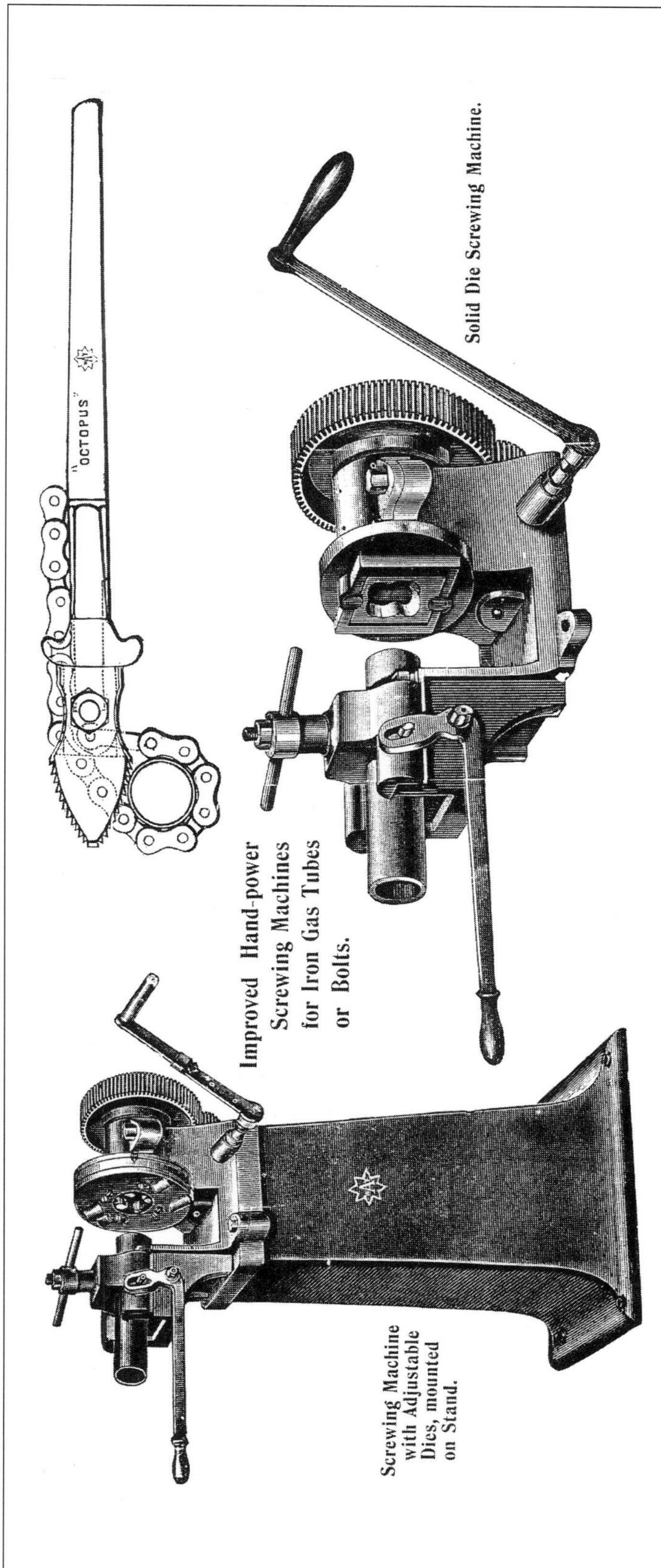

CONTRACTORS TO H M GOVERNMENT AND THE INDIAN AND COLONIAL OFFICES

ALLDAYS, PHONE, BIRMINGHAM."
VICTORIA 2251-4, BIRMINGHAM.

London Office:
9, VICTORIA STREET, S.W.1.
Telephone: 6137 VICTORIA.
Telegrams: "TYPHOON, SOWEST, LONDON."

NATIONAL SCHEME FOR DISABLED MEN

ESTABLISHED · IN · 1650.

ALLDAYS & ONIONS LTD

INCORPORATING
THWAITES BROS. LTD. SAMUELSONS BLOWERS · PETER PILKINGTON LTD.
NEALS EDGE RUNNER MILLS · EADES HOISTS LTD.
NEW ALLDAYS & ONIONS LTD. THE BRITANNIA FOUNDRY CO. LTD.

ENGINEERS
GREAT WESTERN WORKS
SMALL HEATH
BIRMINGHAM 11.

YOUR REF.......................
OUR REF.......... JE/MT.

Directors
A. F. PROCTER, C.A.
R. C. MITCHELL COTTS.
A. HAMILTON.
E. PRICE-JONES.

20th July, 1934.

Manufacturers of

FOUNDRY EQUIPMENT
CUPOLAS.
STEEL CONVERTORS.
LADLES.
ROOTS BLOWERS.
MOULDING MACHINES.
SAND MILLS.
CHARGING MACHINES.
BELLOWS · ETC.

SMITHY EQUIPMENT
PNEUMATIC HAMMERS.
STEAM HAMMERS.
RIVETTING HAMMERS.
PLANISHING HAMMERS.
FORGING FURNACES.
SMITHS' HEARTHS.
ANVILS. VICES.
FORGES, TOOLS, ETC.

FANS
MULTIVANE.
DUST.
PROPELLER.
HIGH PRESSURE.
ETC.

FURNACES
ANNEALING.
NORMALISING.
TEMPERING.
MELTING. ETC.
GAS. OIL AND SOLID FUEL.

SPECIAL PURPOSE PLANT.

CONTRACTORS TO H. M. GOVERNMENT AND THE INDIAN AND COLONIAL OFFICES.

Telephone:
ALLDAYS, PHONE, BIRMINGHAM."
VICTORIA 2251-4, BIRMINGHAM.

London Office:
9, VICTORIA STREET, S.W.1.
Telephone: 6137 VICTORIA.
Telegrams: "TYPHOON, SOWEST, LONDON."

NATIONAL SCHEME FOR DISABLED MEN

ESTABLISHED · IN · 1650.

ALLDAYS & ONIONS LTD

INCORPORATING
THWAITES BROS. LTD. SAMUELSONS BLOWERS · PETER PILKINGTON LTD.
NEALS EDGE RUNNER MILLS · EADES HOISTS LTD.
NEW ALLDAYS & ONIONS LTD. ALLDAYS MOTOR REPAIRS LTD.

SPECIALISTS IN
SMITHY · AND · FOUNDRY PLANT
GREAT WESTERN WORKS
SMALL HEATH
BIRMINGHAM

YOUR REF.......................
OUR REF.......... JL/FP.

Directors:
SIR W. D. MITCHELL COTTS, Bart. K.B.E.
A. T. COCKING, J.P.

QUOTATION. 24. 6. 31.

Contractors to H.M. Government and the Indian and Colonial Offices

Telegrams: "ALLDAYS, BIRMINGHAM"
Telephone No. 487 VICTORIA, BIRMINGHAM

ESTABLISHED IN 1650

London Office: 9 VICTORIA STREET, S.W.1
Telephone No. 6137 VICTORIA
Telegrams: "TYPHOON, SOWEST, LONDON"

ALLDAYS & ONIONS LIMITED

WITH WHICH IS INCORPORATED THE BUSINESS OF THWAITES BROS. (FORMERLY OF BRADFORD)

(ALSO NEW ALLDAYS & ONIONS, ALLDAYS MOTOR REPAIRS, NEAL'S EDGE RUNNER MILLS AND EADES HOISTS)

Specialists in Smithy and Foundry Plant

Directors:
SIR W. D. MITCHELL COTTS, BART., K.B.E.
A. T. COCKING, J.P.

GREAT WESTERN WORKS
SMALL HEATH · BIRMINGHAM

YOUR REF.
OUR REF. WLG/DGB.

11th October, 1927.

Messrs. Wm. Allday & Co.,
38. Paradise Street,
BIRMINGHAM.

Dear Sirs,

ALLDAYS & ONIONS LTD.

On July 18th 1927 the company name changed yet again to Alldays & Onions Ltd., with a nominal capital of £1,000 in £1 shares. This abbreviated name had been used by the company as early as 1905 in some of its catalogues, but now it had risen, almost from the dead, to become the official company title. With the abandonment of motor vehicle and motor cycle production the Great Western Works settled down to the manufacture of its engineering lines, although some cycle work continued, mainly in tradesmen's cycles and tricycles, until about 1935.

By March 1929 the financial position of the company had improved to the extent that expansion of the factory took place in the form of a new Light Machine Shop which was built partly over the open yard between some of the existing buildings which added 4,000 square feet of production space. In October of the same year a large 3 bay extension to the Foundry was erected, part of this extension being built on land which had previously been owned by Robert Coles & Co., and latterly by J.H.King, a local coal merchant, taking the Alldays works right up to the back gardens of the houses at the corner of Anderton Road and Sydenham Road. The new foundry extension covered some 7,800 square feet and had required the demolition of some of the old buildings.

In 1931 Mr R. Crichton Mitchell Cotts replaced his father as a Director of the company and just about one year later Sir William Mitchell Cotts died. Mr. A. Hamilton then joined the Board of Directors and, after the retirement of Mr A.T. Cocking shortly afterwards, was joined by Mr L. Aldridge with Mr. E. Price-Jones as Managing Director. (Mr Price-Jones had previously worked at the Sentinel Works in Shrewsbury.)

A gradual rationalisation of the product range took place with Thwaites designed steam hammers, copulas and Roots blowers replacing the Alldays products and Pilkington pneumatic hammers replacing the Alldays range with the exception of riveting and planishing hammers.

Several types of bellows were still in production including house bellows in sizes from 8" to 18" which cost £1-16s-0d to £6-15s-0d each in standard quality, or £3 to £10 in best quality red bound and quilted models with brass pipes and nails. Electricians bellows were available from 8" to 15" from 16s-0d to £1-5s-6d each and much larger Smiths bellows in sizes from 18" to 42" which cost from £1-18s-0d up to £11. Lever operated double blast circular bellows were also still available in sizes from 18" diameter up to 42" in diameter and cost £5-9s-3d up to £22-15s-0d each.

The 'LS' range of "Silent Fans" introduced a few years earlier had been increased to include models from 14" diameter up to 24" in diameter with direct drive by electric motor having superceded belt driven models, although this could still be specified for the smallest model then available. These fans covered the range of 100 cu.ft/min up to 2000 cu.ft./min and the smaller types had cast iron casings with die formed impellers, whilst larger models had fabricated mild steel welded casings.

The improved "Serlano" dust fans had fabricated paddle blade type impellers running in fabricated sheet steel casings and were

ILLUSTRATION

Interesting changes in the company letterheads are shown here, between 1927, just after the re-formation of the company as Alldays & Onions Ltd., which was just two years after the formation of New Alldays & Onions Ltd., and 1934.

The 1934 letterhead gives some idea of the enormous range of products still being made by the company as these are listed down the left hand side of the page.

6 Q TONS
THWAITES BROS LTD
BRADFORD 1908

M
R

PHOTOGRAPHS

LEFT-TOP: A 60 ton capacity bottom pouring geared ladle made to Thwaites design for use in the largest of iron foundries around the world. A 2½ ton ladle weighed 16 cwts, so you can imagine the weight of one of these monsters.

LEFT-BOTTOM: This interesting picture shows a consignment of eight 64 ton geared ladles being transported to the docks by rail for shipment to Japan.

BELOW: This photograph was taken at the junction of Sydenham Road and Anderton Road, just a few yards from the works, and shows an Alldays cycle coupled to a cycle trailer. The trailer was patented by the company in October 1930, the two patentees being Leigh Evans Clulee and Victor Earnest Barker. The trailer was so designed as to allow the cycle to be 'banked' on cornering, without lifting the trailer wheels. This was accomplished quite simply by pivoting the triangular frame, used to attach the trailer to the cycle, at the underframe of the trailer. When disconnected from the cycle the triangular frame could be used to push the trailer around as a hand truck. The load carrying box on the trailer was provided with five coil springs as a form of suspension between it and the underframe, and this ingenious little trailer cost £8-10s-0d.

available in sizes from 12" to 54" diameter. These were available as belt driven or electric motor driven models.

The "Climax" was a new range of multivane fans with three different designs of fabricated impellers, the 'CF', 'CR' and 'CB' type. The casings were made from fabricated steel sheet welded together, but in the smaller models 9" to 36" the casing was bolted between two cast iron side plates. In sizes 42" to 60" the fabricated casings were split horizontally at the centre line of the impeller drive shaft. All models could be specified for belt drive, direct or indirect electric motor drive, or by an engine. Fans with a 'CF' type impeller covered a range from 1500 cu.ft/min up to 100,000 cu.ft/min, with a 'CR' type impeller from 1000 cu.ft/min up to 100,000 cu.ft/min and with a 'CB' impeller from 1000 cu.ft/min up to 80,000 cu.ft/min.

This range was also complimented by double inlet double width fans, D.I.D.W. types, which were cheaper than a comparable single inlet fan, required less headroom and used a smaller motor to drive it. Belt driven single inlet fans cost from £12-10s-0d up to £155 and weighed 140 lbs for a number 9 up to 6,600 lbs for a number 60. Larger fans than the standard range offered could also be made to customers' own specifications.

New foundry moulding machines were introduced under the "Coventry" and "Britannia" trade names for producing sand moulds on a production line basis. Five pneumatically operated models formed the "Britannia" jar-ram range which could handle moulds from 5 cwts up to 30 cwts, the smallest model having a table 1ft 9ins by 1ft 5ins and the largest model a table 4ft 6ins square. A smaller hand operated moulding press was available in this range suitable for very shallow castings which, although only operated by a hand lever produced about 1 ton of pressure.

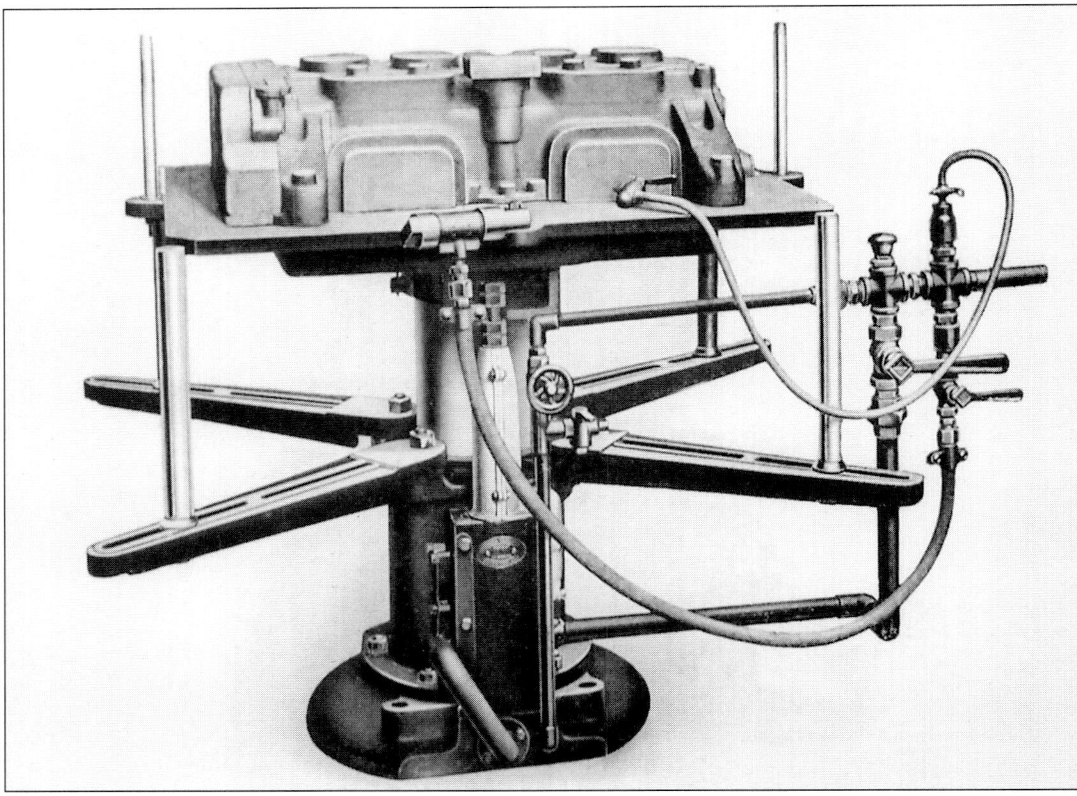

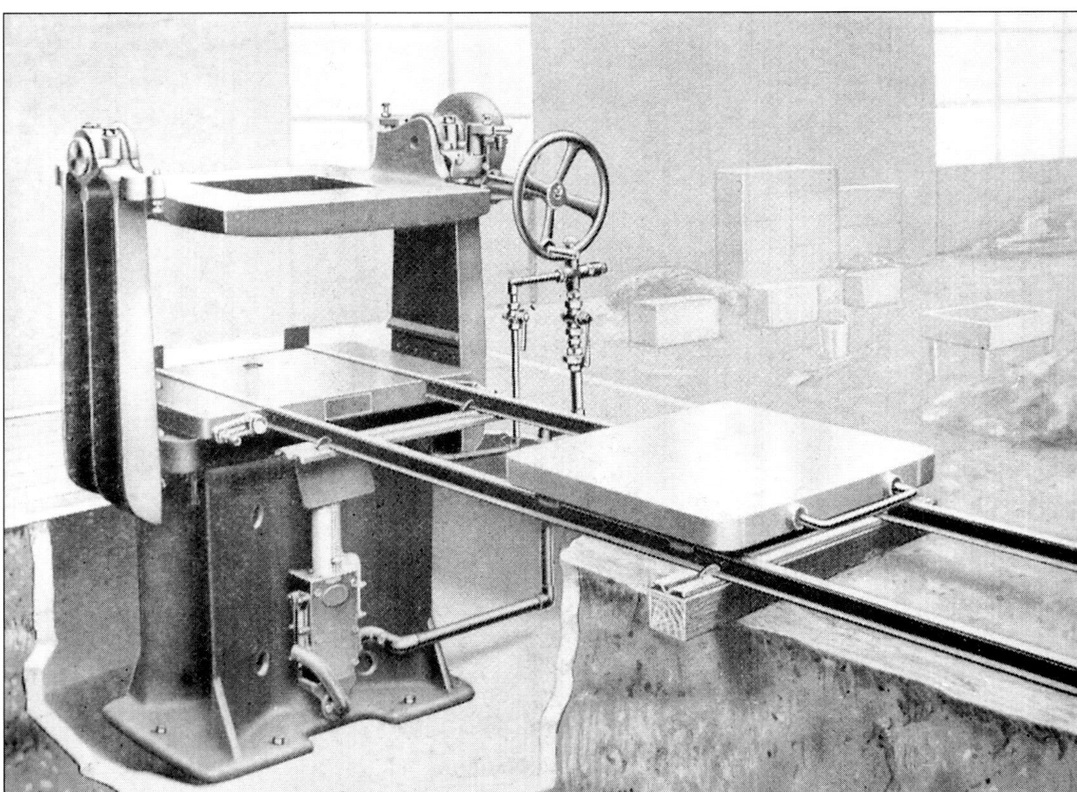

PHOTOGRAPHS

LEFT: With an extensive foundry of its own making all sizes of castings for its own products the company was able to apply considerable knowledge and 'hands on' experience to produce an extensive range of moulding machines suitable for both large and small foundries.

Jar-Ram Moulding Machines, to make the sand mouldings, were made under the "Coventry" and "Britannia" brand names.

The top picture shows a "Britannia" No. 1 machine with an engine crankcase pattern plate in position on a casting box 2ft 3ins by 4ft. This pneumatically operated machine was capable of ram moulding up to 10 cwts and weighed about 10½ cwts.

The lower picture shows a "Coventry" No. 1 machine in a typical setting as installed in a large foundry producing castings in large numbers. This machine featured a top turn-over table operated by the handwheel with the ram table itself pneumatically operated by the large piston visible beneath the machine. Available in four different sizes these machines were able to cope with moulds from 8 cwts up to 30 cwts and the machine illustrated measured 3ft 6ins long by 2ft 4ins wide, stood 5ft 3ins high and weighed 25 cwts.

BELOW: This "Empire" oil fired forging furnace with fan blast had a heating space 29ins by 18ins, could heat bars up to 2½ins square, used oil at the rate of 2½ gallons per hour and cost £58, less the firebrick lining. A range of 12 models was produced costing up to £95.

The "Coventry" range included a hand operated moulding machine with a table 1ft 8ins by 1ft 10ins and constructed to enable a pressure of up to 5 tons to be applied to the mould. The other five machines were of the pneumatically operated type with a turn-over facility for the mould and provided the means for high volume production of foundry moulds from 8 cwts up to 30 cwts. A pneumatically operated sand riddle machine was also included in the "Coventry" range of products which could riddle, or sift, sand as fast as a man could shovel into it! Bottom pouring geared ladles which prevented slag from contaminating castings, of Thwaites design, for steel foundries were made in a standard range of capacities from 10 cwts to 50 cwts which cost £28 to £57, but ladles up to 60 tons had also been produced in large numbers and some monsters of 64 tons capacity had recently been made for Japan! At the other end of the scale hand ladles were still being made with a capacity of 28lbs which cost 16s-6d each. Top pouring geared foundry ladles were also available in sizes from 10 cwts up to 4 tons and cost £18 to £52.

Furnaces for various uses were being made under the "Empire" trade name and these included gas or oil fired hardening and annealing furnaces which could operate up to 1020°C, oil fired forging furnaces and oil fired rivet heating furnaces, both static and portable models. Also made under the "Empire" name were oil or gas fired crucible tilting furnaces with capacities of 40 lbs to 500 lbs in the standard range and up to 1,000 lbs to special orders. These oil fired furnaces cost £50 to £97-10s-0d in the standard range. "Empire" steel tyre furnaces were also still in production for heating railway locomotive/wagon tyres from 20" diameter to 72" diameter.

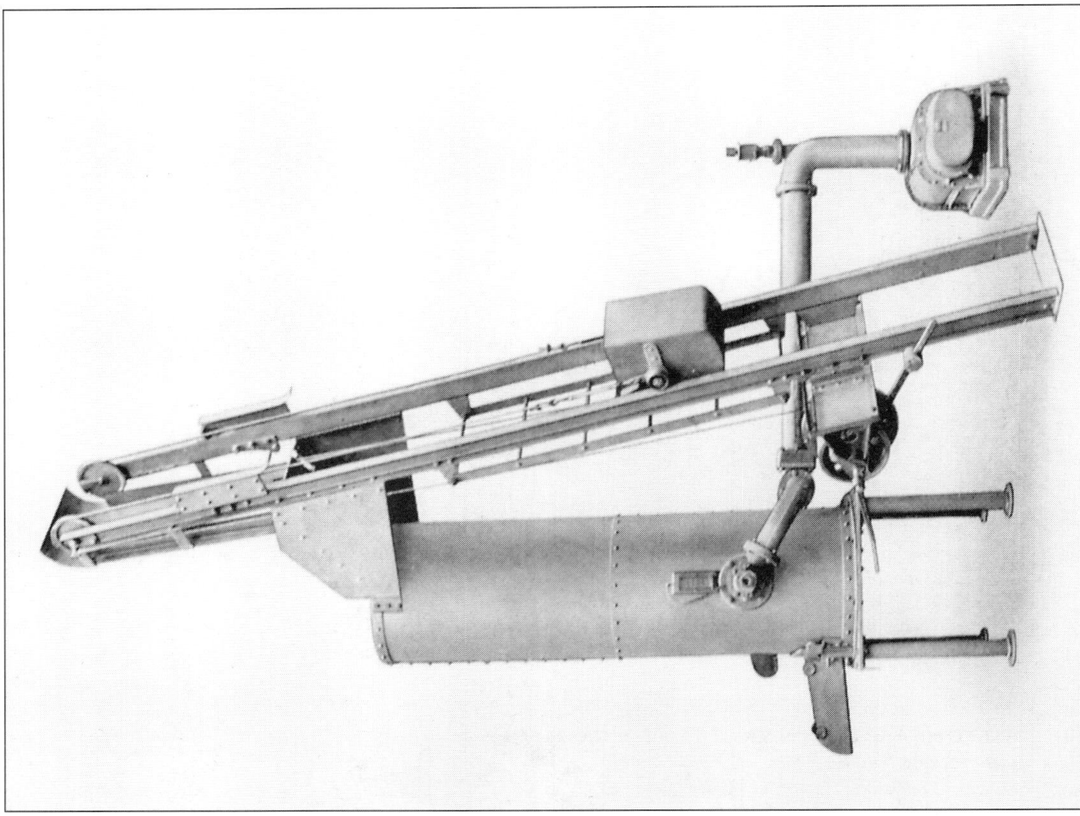

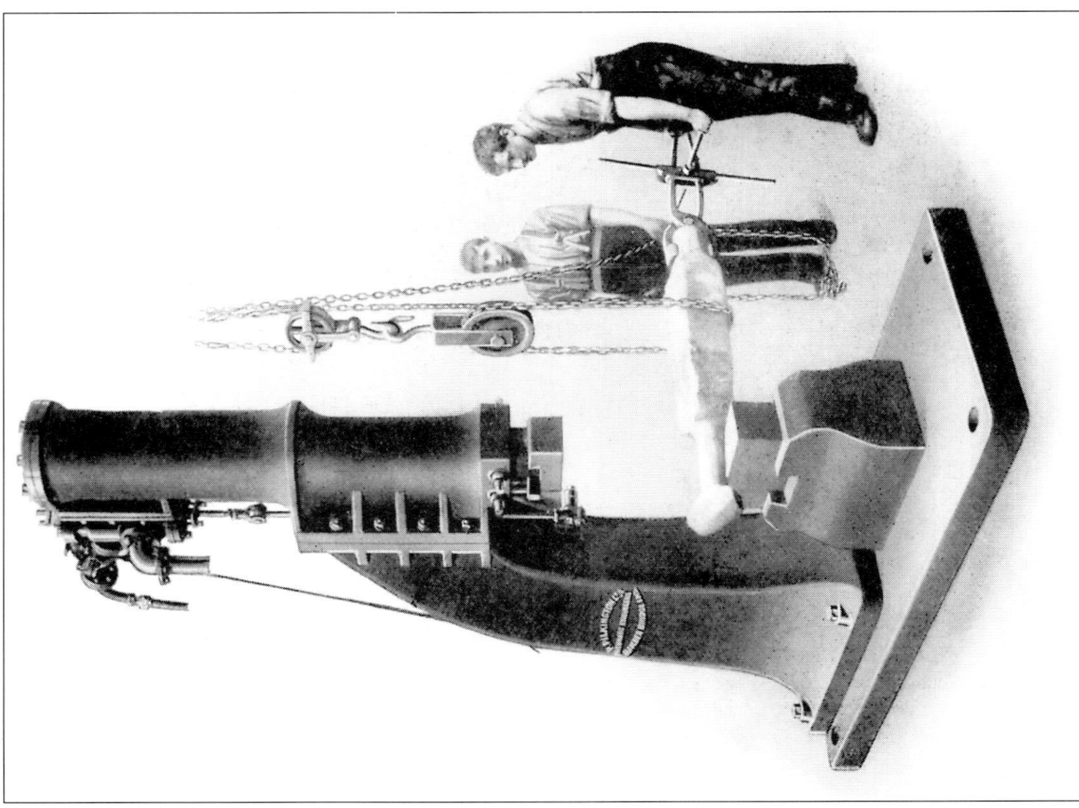

PHOTOGRAPHS

LEFT-TOP: This photograph shows two of the company's products, a Dwarf Cupola together with a hand operated Charging Machine. The charging machine featured a hand operated winch which raised and lowered the bucket via wire ropes, the bucket automatically being discharged into the cupola. One man could operate the machine which could be made portable to feed more than one cupola. Peter Thomas who served his apprenticeship with the company recalled one huge charging machine being made which required a hole to be knocked through the workshop floor into the basement, to enable it to be assembled and tested in the works before being despatched to the customer. The cupola itself is shown here connected to a Roots type blower.

LEFT-BOTTOM: A 5 cwts "N.S.K." Compound Pneumatic Hammer which was made in a range of 1 to 40 cwts. This improved range of hammers was claimed to be more economical to operate due to the patented expansion gear which could be supplied as a kit to modify existing hammers without having to buy a new standard baseplate or anvil. The 5 cwts hammer shown here could draw out a 7" cube of steel into a 2" square bar 6' 6" long in one heat. These hammers were used by companies such as, Anthony Gibb & Sons, Melbourne, Burn & Co., Bengal, the British Admiralty, J. Brockhouse & Co.Ltd., West Bromwich, Cammell Laird & Co.Ltd., Birkenhead, Delaware & Lackawanna Railway, U.S.A., Fodens Ltd., Sandbach, Harland & Wolfe Ltd., Belfast, Italian State Railways, Vickers Ltd., Barrow in Furness, Leyland Motors Ltd., and the LMS Railway Company, Crewe.

ILLUSTRATION

RIGHT: An Ultra Rapid Cupola with drop bottom, without a receiver and with much of the chimney removed. This range of cupolas had melting capacities from 1-2 tons up to 12-15 tons. The smallest cupola was 2 feet 6 inches in diameter, stood just over 22 feet high, weighed 2 tons less its fire brick lining and cost £70. The largest machine was 8 feet in diameter, stood 40 feet high, weighed 16 tons 10 cwts, held 36 tons of fire bricks and cost £420.

2½/19 L.S. Motor Driven Fan.

Size 2 L.S. Driving Side.

PHOTOGRAPHS

LEFT: The 'LS' range of fans had been developed, initially, to provide small volumes of air at fairly high pressures, with the added advantage of silent running. Fans made in the 2 LS 14" to 19" range were made with a cast iron casing split vertically and were available with electric motor drive, or belt drive. With a motor speed of 2,900 r.p.m. these fans produced 30-150 cu.ft/min.

The larger models in the 'LS' range had casings and motor support stools made from fabricated mild steel plate which was of welded construction and these were available in sizes from 14" to 27" and with outputs of 60 cu.ft/min up to 2,100 cu.ft/min. These larger sizes, which were designated 2½, 3, 3½, 4, 4½, 5 and 6, were also available driven directly by electric motors or with belt drive.

This range of silent fans was to remain in the company's range of products into the 1970s, although by then many other sizes had been developed and made available.

ILLUSTRATIONS

RIGHT: Forges of all sizes were still an integral part of the new company's catalogue of products and these illustrations show a small selection of forges which could be used in the workshop, or as portable units for outdoor use.

From top to bottom the illustrations show the "Climax" with Roots blower and bottom blast tuyere, the "Climax" patterns A330-2 and patterns A325-7 and the "Ajax" HF817 model.

The first "Climax" model was available in three hearth sizes, 20", 22" and 24" diameter, weighed 200 lbs, 260 lbs and 280 lbs respectively and cost £11-5s-0d, £12-14s-0d and £13-19s-6d.

The other two "Climax" forges featured a hand geared Roots blower with the air blast directed into the hearth horizontally from the back of the hearth. The patterns A330-2 models were also made in sizes with hearths 20", 22" and 24" diameter and cost £ 11-11s-0d up to £13-15s-6d. The patterns A325-7 were made with hearths 20" by 26", 24" by 30" and 30" by 36" and cost £12-5s-0d up to £17-10s-0d.

The "Ajax" featured a 12" diameter gear driven fan with water cooled tuyere and cost £5-5s-0d.

" Climax " Forge.

Patterns A330–2.

Patterns A325-7.

The " Ajax " Forge, HF817.

Dimensions of D.I.D.W. Fans

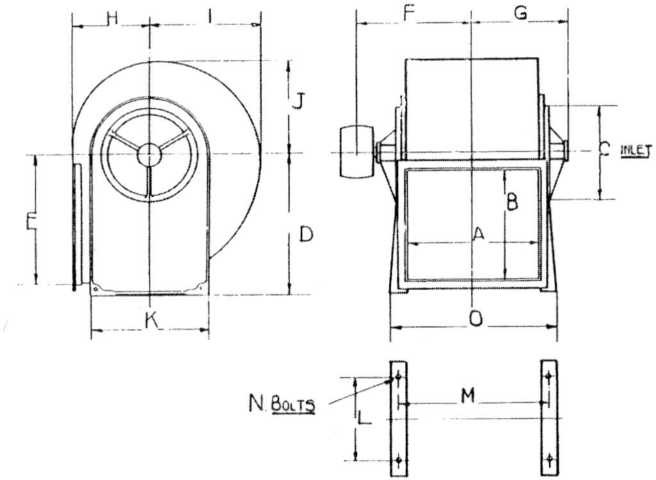

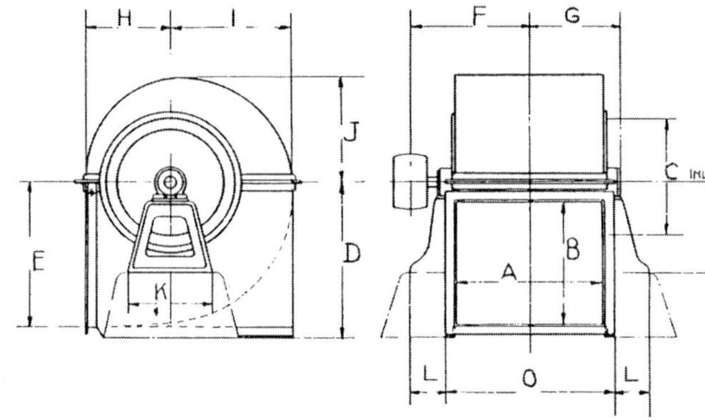

Fan No.	A	B	C	D	E	F	G	H	I	J	K	L	M	N	O	Pulley Dia.	Pulley Face
12	17¼	14¾	14½	19	16⅞	16	13⅝	10	14½	12 1/16	17	12	19⅞	½	21½	7½	4½
15	21½	18½	18	23	21	18½	15½	12	18 1/16	15 1/16	19	14	24⅛	½	25¾	9	5½
18	25¾	22	21¼	27½	25¼	21½	18⅛	14⅜	21⅝	18 1/16	24	17½	29⅛	¾	31¾	10½	6½
21	30	26	25¼	32	29½	24½	20⅞	16⅞	25¼	21⅛	28	21½	33⅛	⅝	35½	12	7
24	34½	29½	28¾	37	33⅝	27	23⅜	19¼	28¾	24⅛	33	26	38	⅝	40¾	13½	7
27	38¾	33	32¾	41	37¾	30¼	25½	21⅝	32½	27⅛	37	28	42	⅝	44¾	15	9

Fan No.	A	B	C	D	E	F	G	H	I	J	K	L	M	O	Pulley Dia.	Pulley Face
30	43	37	36	45½	42	35	26½	24	36½	30	25	10	26	49¾	18	10
36	51½	44	43	54	50½	40	30¾	31	43½	36	30	11	33	57¾	21	12
42	60	53	50½	62½	59	45	36	33¾	50½	42	35	12	36	67⅜	24	12
48	69	59	57½	71	67¼	50	40½	38½	57¾	48	40	13	41	76⅜	27	12
54	77½	66	65½	79½	75½	55	45¾	43	64½	54	45	14	46	86½	30	12
60	86	74	72	88	84	60	50	48	71¾	60	50	15	57	94½	33	12

DIAGRAM

LEFT: This diagram and table gives all the dimensional data for the range of double inlet double width (D.I.D.W.) range of fans. The largest fan in this range had a diameter of 5 feet, stood 12 feet 4 inches high, was over 9 feet wide and had an outlet measuring 6 feet 2 inches by 7 feet 2 inches and was driven by a pulley 2 feet 9 inches in diameter and 1 foot wide.

ILLUSTRATIONS

RIGHT: These illustrations show the various types of multivane impellers developed for the "Climax" range of fans, together with an illustration of a standard belt driven model made in sizes from 42" diameter up to 60" in diameter with fabricated mild steel plate casings and with the belt driven shaft mounted on a cast iron baseplate independent of the fan casing itself.

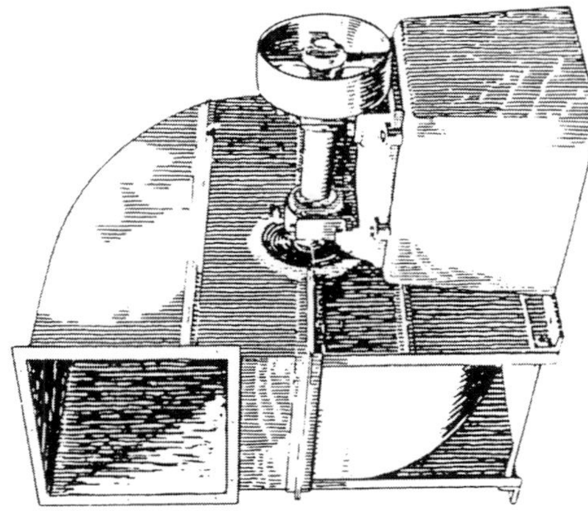

"C.B." Type

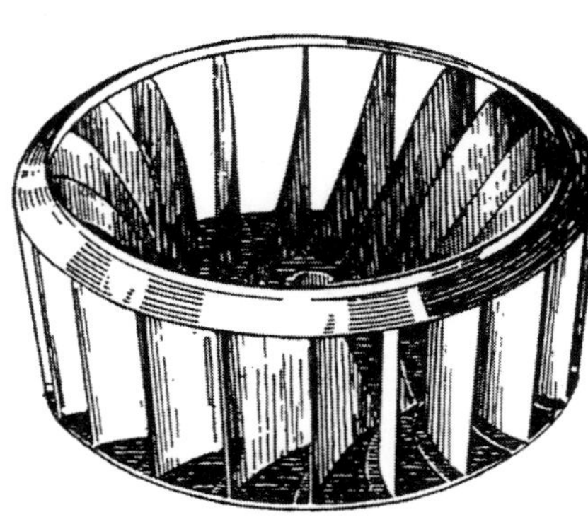

"C.R." Type

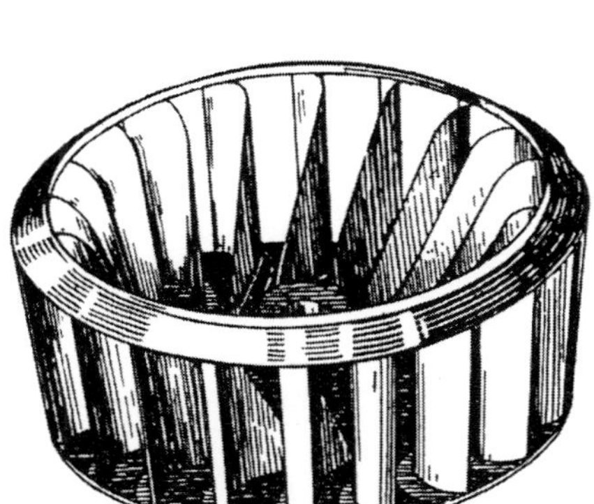

"C.F." Type

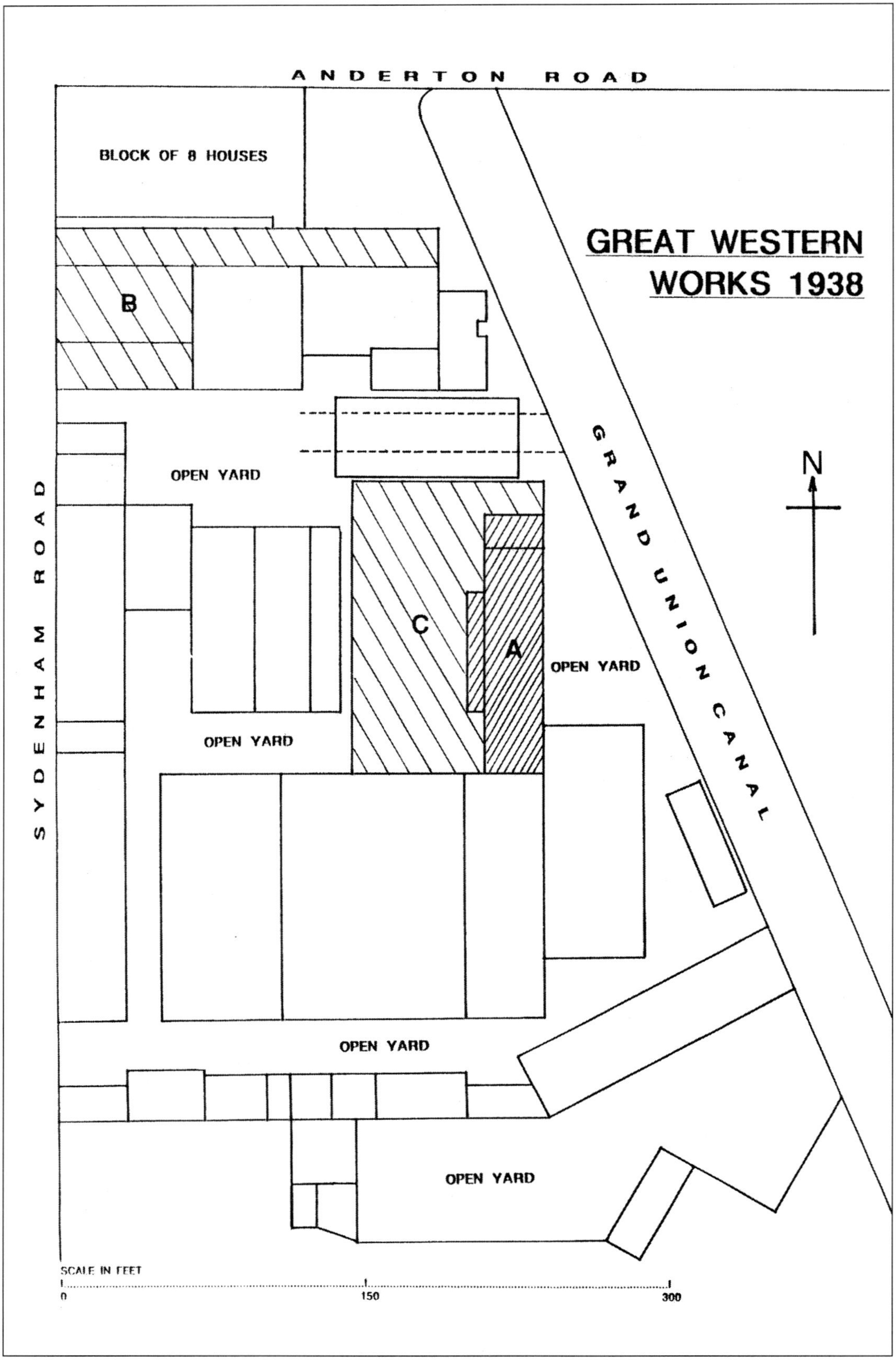

ANDERTON ROAD
BLOCK OF 8 HOUSES
GREAT WESTERN WORKS 1938
B
SYDENHAM ROAD
OPEN YARD
GRAND UNION CANAL
N
C
A
OPEN YARD
OPEN YARD
OPEN YARD
OPEN YARD
SCALE IN FEET
0
150
300

BUSINESS AS USUAL INTO WARTIME

As early as February 1937 the company began supplying equipment to produce war material when a Shell Nose Furnace was put back into production. This furnace was followed in 1939 and 1940 by furnaces for 5", 6", 9.2" and 75mm shell noses. Winches of various types had been part of the standard items available from the company from about 1900 and a new 5 ton winch was introduced in 1941, together with special boat hauling winches for the Admiralty which were tested on a rig driven by a Morris petrol engine.

In 1938/39 the factory had been enlarged by the erection of a new Heavy Machine Shop which was equipped with the latest electrically driven machinery for the production of cast iron cased fans and power hammers.

Much of the work being carried out at Sydenham Road during the war was for the Ministry of Supply which was responsible for equipping factories around the country with machinery, but contracts were also undertaken for the Admiralty and Ministry of Aviation. About 55% of the company's wartime production was for the Ministry of Supply with 40% going to the Admiralty and the remainder to the Ministry of Aviation. Parts for a supercharger were being manufactured from 1939, possibly for an aircraft engine and one item listed in an old Pattern Record Book for 1943 was for 'Whittle Wheels' which could, possibly, have been parts for the then top secret jet engine. Other items described as road wheels of 14" and 8" diameter may well have been parts for tracked armoured vehicles.

Cast iron marking off tables and surface/fitting plates for the engineering industries were produced in abundance, the marking off tables being made in sizes from 3' x 2' up to 5' x 10'. A 3' x 2' table weighed 7 cwts and cost £10-16s-9d and a 6' x 4' table weighed 18.5 cwts and cost £30-10s-0d. Circular surface plates had been in the Alldays catalogue since 1905 and were made in sizes from 6" in diameter up to 24" diameter. High grade surface plates made from close grained cast iron hand finished to an accuracy of 0.0003" per foot were made in sizes from 4" x 6" up to 36" x 36". These weighed 9 lbs to 545 lbs, cost from 13s-6d up to £21 respectively and were supplied with oak covers to protect the working surface from damage and dust. Other special sizes were made during the war up to 6' x 9'.

Levelling Slabs for blacksmiths and anglesmiths were also made from cast iron 2.5" thick with sides 6" deep and in sizes from 3' x 2' which weighed 8.5 cwts and cost £10-5s-6d up to 6' x 4' which weighed 30 cwts and cost £36-13-6d.

Vice benches and work benches, made under the "Anchor" trade name, were also produced in cast iron from 3' x 4' up to 6' x 15', although the standard range normally only extended to 4' x 9', these weighed 12 cwts to 35 cwts and cost £16 to £42 respectively. A similar range of fitters' benches was also made in sizes up to 4' x 12' with tops of seasoned elm. Cast iron bench legs were also available separately in widths from 24" to 60" for customers wishing to build their own benches and parts were made for the large drawing boards used by draughtsmen. Castings were also made for other companies such as Dunlop tyre moulds and Higgs Motors who made electric motors in their Witton, Birmingham, factory.

The factory was exceptionally busy trying to supply smithy plant, power hammers and furnaces of all types as fast as they could be made to equip factories around the country involved in war work and for one period employees were working 7 days averaging 70 hours per week.

DIAGRAM

1938 BLOCK PLAN OF THE "GREAT WESTERN WORKS"

A. Light Machine Shop 112 feet long by 28 feet wide, 20 feet 7 inches high to the eaves and 28 feet 6 inches high overall. A 10 ton overhead crane was mounted at a height of 16 feet 3 inches above the workshop floor, running the whole length of the building. Three 24" diameter 'Robertson' ventilators were mounted on the roof. A smaller two storey building was erected along the west side of the main workshop, 60 feet long by 8 feet 6 inches wide and 23 feet high to the ridge. This smaller building was accessed via a steel staircase within the main workshop and the upper floor of this building was 11 feet above ground level. At the north end of the main workshop was a small 'lean-to' type of building 28 feet by 16 feet, 6 feet 9 inches high to the eaves and 13 feet 4 inches high overall. All three buildings were erected in 1929 by Nortons (Tivedale) Ltd.

B. This extension to the factory, which was also built in 1929 by Nortons (Tivedale) Ltd., comprised three bays, necessitating the removal of some existing buildings and also took in land previously owned by J.H.King, coal merchant, at the north end of the site. The longest bay was 185 feet 6 inches long by 22 feet 4 inches wide, 22 feet high to the eaves and 28 feet high to the ridge. This bay was designed to house a 3 ton overhead crane mounted 16 feet 9 inches above the workshop floor.
The second, and largest of the three bays, was 65 feet long by 45 feet 3 inches wide, 23 feet high to the eaves and 34 feet 6 inches high overall. Seven 30" diameter 'Robertson' ventilators were fitted in the roof of this bay adding a further 4 feet 9 inches to the height of the building. This bay was also designed to house an overhead crane mounted 15 feet 9 inches above the workshop floor. The third bay of this extension was 65 feet long by 15 feet 4 inches wide, 17 feet 9 inches high to the eaves and 22 feet high to the ridge. A 2 ton, hand operated, overhead crane was mounted in this bay 11 feet 6 inches above the workshop floor.

C. This 'L' shaped Heavy Machine Shop was designed by Ewan Harper, Bros & Co. in 1938 and was 135 feet long by 57 feet wide, 29 feet high to the eaves and 37 feet high to the ridges of the six asymmetric roof spans which incorporated Northern Lights. The small leg of this 'L' shaped building replaced the small 'lean-to' type structure at the north end of the Light Machine Shop, building 'A'.

PHOTOGRAPHS

LEFT: The top picture shows a Pattern B 2913 geared power press suitable for trimming stampings and could cater for stampings up to 1 ton in weight. The press stood 7ft 2ins high with a base measuring 30ins by 35ins, had a flywheel 42ins in diameter, developed a pressure of 95 tons, and weighed 3 tons 6 cwts. The price was £197.

The picture below shows a riveting machine developed during the First World War to rivet together the tracks for military tanks, and afterwards offered to customers for general engineering work.

ILLUSTRATION

RIGHT: These two pictures show an "Empire" sand mixer and disintegrator designed for use in foundries to grind and mix sand for moulds, but could also be used for many other materials.

Figure 1 shows the machine complete with casing over the reel and with the drive belt pulley to the left.

Figure 2 shows the machine with its casing removed to reveal the adjustable reel through which the sand was passed for grinding and mixing.

Four sizes of machines were offered with capacities of 1 to 6 tons per hour, requiring from 3 h.p. to 10 h.p. to drive and weighing from 5 cwts to 16 cwts.

Fig. 2.

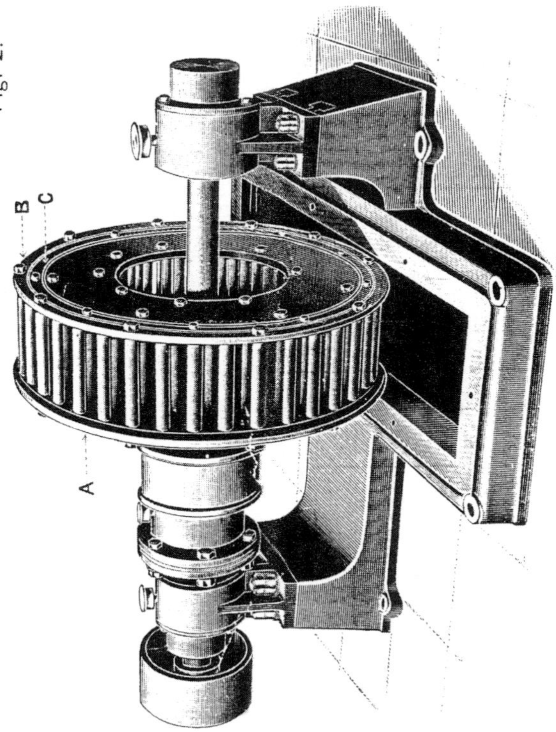

Fig. 1.

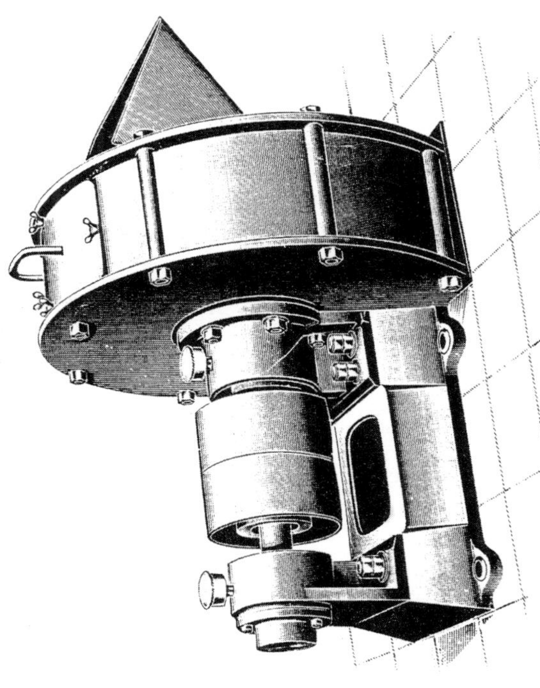

On October 22nd 1940 a 500 lbs bomb landed just outside the new heavy machine shop making a crater 50 feet across and doing considerable damage to the Smithy Shop and part of the machine shop. Even before this damage had been repaired the works was hit by incendiary bombs which destroyed the fitting shop office, destroying many valuable company records.

On December 2nd 1940 the three storey building close to the canal was also hit by incendiary bombs setting fire to the pattern and

PAGE 3

"SERLANO" DUST FANS.

DIMENSIONS.

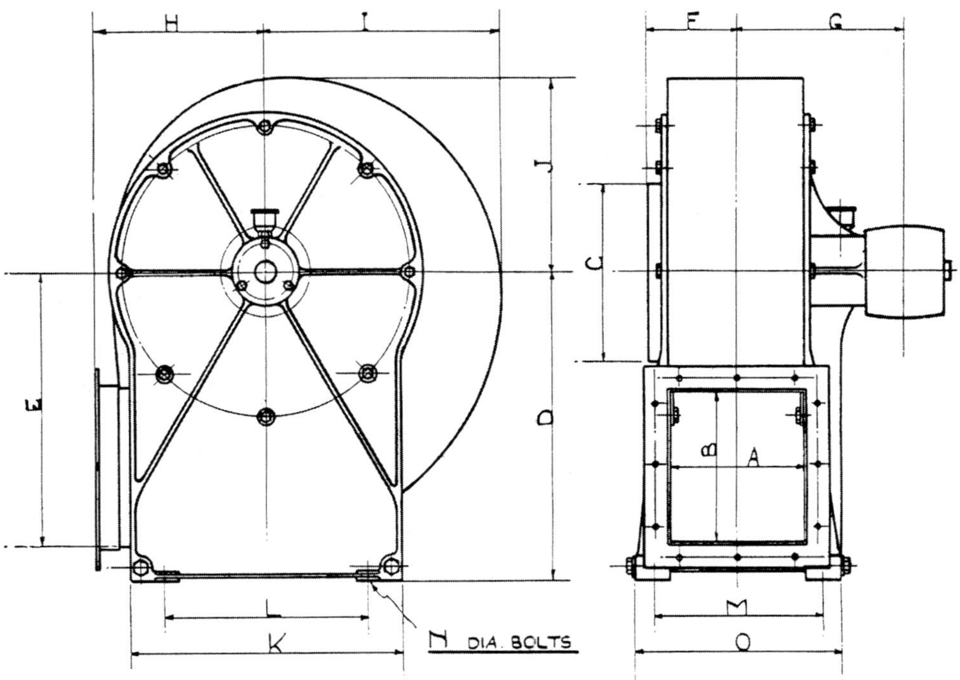

Fan No.	A	B	C	D	E	F	G	H	I	J	K	L	M	N	O	Pulley.	
																Dia.	Width.
12	6″	6¾″	7½″	13½″	12″	4″	7½″	7½″	10¼″	8½″	12″	9″	7⅞″	⅜″	9⅛″	4″	3½″
15	7½″	8½″	9½″	16¾″	15″	5″	9½″	9½″	12⅞″	10¾″	15″	11⅛″	9½″	⅜″	11⅜″	5″	4″
18	9″	10″	11¼″	20″	18″	6¼″	11¼″	11¼″	15½″	13″	18″	13″	11¼″	½″	13¼″	6″	5″
21	10½″	11¾″	13″	23″	21″	7¼″	13″	13″	18″	15″	21″	16″	13″	½″	16⅝″	7″	6″
24	12″	13½″	15″	26″	24″	8″	15″	15″	20½″	17″	24″	18″	15″	½″	18⅛″	8″	7″
27	13½″	15¼″	17″	29½″	27″	10″	17″	17″	23¼″	19¼″	27″	20¼″	17″	½″	19¾″	9″	7″
30	15″	17″	19″	32½″	30″	10⅜″	19″	19″	25¾″	21¼″	30″	22½″	18½″	⅝″	21¼″	10″	8″
36	18″	20¼″	22½″	38½″	36″	12¼″	22½″	22½″	30⅞″	25½″	36″	27½″	22½″	⅝″	24½″	12″	10″
42	21″	23½″	26″	45″	42″	14″	26″	26″	36″	30″	42″	32″	26″	⅝″	27½″	14″	12″
48	24″	27″	30″	52″	48″	15¼″	30″	30″	41″	34¼″	48″	36½″	30″	¾″	32¼″	16″	14″
54	27″	30½″	34″	60″	54″	18″	32″	34″	46½″	38½″	54″	44″	33½″	⅞″	37½″	18″	14½″

DIRECTIONS OF DISCHARGE.

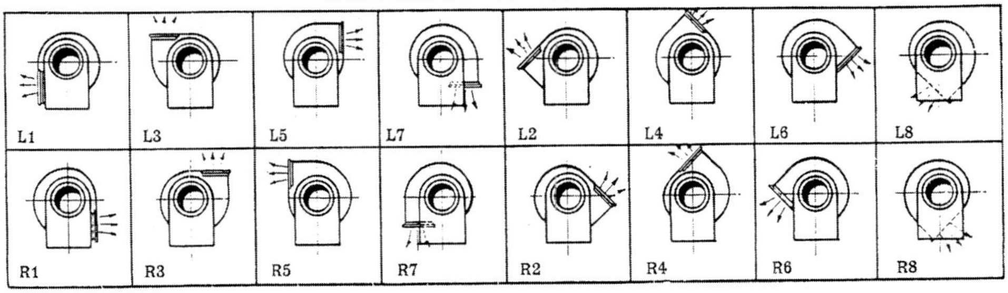

ILLUSTRATION

LEFT: Drawing showing a belt driven 'Serlano' dust extraction fan, together with a table of dimensions and illustrations showing the various positions available for the outlet, or discharge, of the fan to suit different installations. The casing was made up of welded sheet steel bolted between two cast iron side plates. The runner(fan) was of the paddle blade type, the thick blades being riveted to arms cast integrally with the hub. These fans were available with belt drive pulleys, driven directly by electric motor, or through a coupling to an electric motor.

This range of fans had outputs from 960 cu.ft./min up to 22,500 cu.ft./min and cost from £10 for the smallest belt driven model up to £110-10s for the largest electric motor driven model complete with cast iron bedplate.

PHOTOGRAPH

This pneumatic hammer was developed to provide a Shell Riveting Hammer. The picture on the left shows the hammer with shell mounted in the anvil ready for riveting the base plate into a shell. The picture on the right shows the anvil thrown out of gear for loading/unloading the shell, which is shown here in front of the anvil. The shell holder on the anvil was revolved by worm and worm wheel belt driven from the main drive shaft of the hammer itself. A 3.3" shell could be riveted in 20 seconds and a 4.5" high explosive shell in 30 seconds.

bellows shop on the top floor and auxiliary firemen tackling the blaze had to climb over rows of barges tied up at the wharf before they could get their suction pipes into the canal to draw water. Whilst this fire was being tackled a second wave of bombers dropped oil bombs which hit the main office block setting fire to the drawing office. This two storey building which ran along Sydenham Road had wooden floors which soon caught fire which rapidly spread throughout the length of the building, destroying in its wake the millwright's and carpenter's shops on the ground floor. The whole building with the exception of the ground floor strong room at the Anderton Road end of the works was gutted. One worker remembers finding all the medals awarded to the company over many years whilst sifting through the debris, but what became of them afterwards is still a mystery to this day. Although Alldays was obviously making a valuable contribution to the war effort the bombing is thought to have been aimed at the B.S.A. works just across the Golden Hillock Road. Beneath the floor of one of the big machine shops was a cellar which, until the destruction of the office block, had been used as an air raid shelter for employees and some of the people living in Sydenham Road. This was now converted into office accommodation for the accounts, wages, progress department and Works Manager, whilst sales and executive staff were rehoused, temporarily, in the nearby 'Marlborough Hotel' public house situated at the corner of Anderton Road and Montgomery Street.

All the woodworking machinery in the pattern and bellows shop had been destroyed and for a short time work continued by having to make parts by hand. The ability to produce some of the products from memory after the loss of patterns and drawings was almost entirely due to the experienced 'old timers' who had been with the company all their working lives. The works' carpenter set about making tables and stools and the drawing office was rehoused in a room over the works' electricity sub station.

The rest of the war proved less 'eventful', other than the occasional air raid warning during the daytime and the factory output continued without any further interruptions.

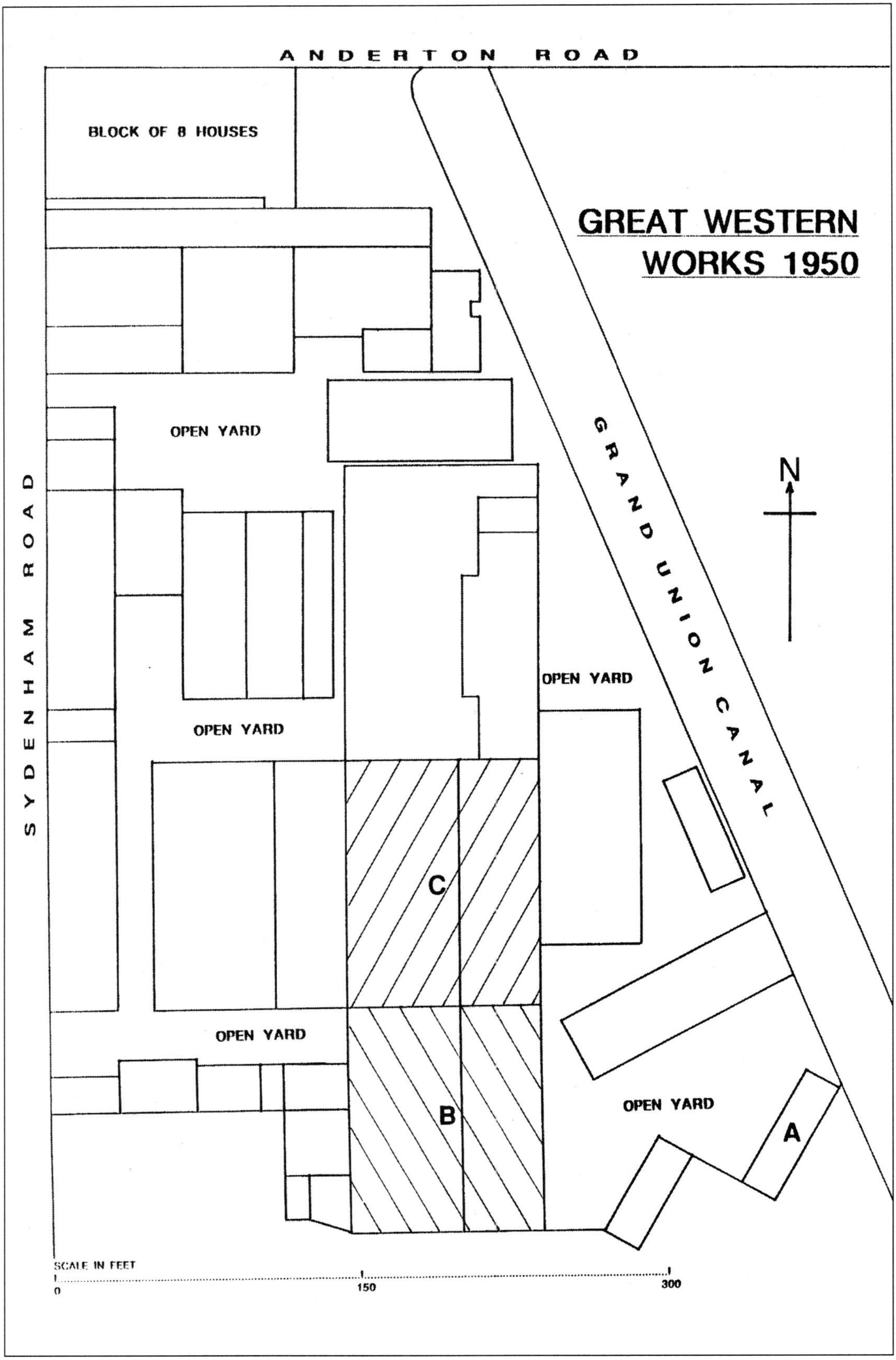
ANDERTON ROAD
BLOCK OF 8 HOUSES
GREAT WESTERN
WORKS 1950
OPEN YARD
SYDENHAM ROAD
GRAND UNION CANAL
N
OPEN YARD
OPEN YARD
C
OPEN YARD
B
OPEN YARD
A
SCALE IN FEET
0
150
300

POST-WAR DEVELOPMENTS

The company's range of post-war products continued much along the lines of pre war days with Duplex, Climax, Waverley and Serlano fans, Forging, Annealing, Rivet Annealing and Tilting Furnaces, Steam Hammers from 5 cwts to 5 tons, Beam Hammers, Pneumatic Hammers, Blowers, Charging Machines and Loam & Sand Mills.

Fortunately, the company found the early post-war years very busy indeed with British industry undergoing a substantial rebuilding programme and it was not long before the need for more production space became pressing, requiring the company to implement its own modernisation programme.

The old self-contained, adjoining buildings were opened up by removing walls to bring together many of the workshops into a common workplace, and further extensions were planned.

In 1948 a single storey Machine Shop measuring 98 feet by 112 feet was built, which required the partial demolition of some smaller buildings which, as a virtual extension to two other existing buildings, resulted in a main workshop measuring 98 feet by 367 feet. The centre section of this workshop had been partly rebuilt in 1915, but in 1950 this was rebuilt again to match the height of the later buildings.

In 1949 Mr Price-Jones, the Managing Director, died after undergoing an emergency operation and was succeeded by Mr. J.Erskine, who retired from the company in 1957 and was replaced by Mr. L. Lee, who had previously been General Manager. Following the death of Mr. L. Aldridge in 1952 his place on the Mitchell-Cotts Board of Directors was taken over by Mr. H.Drayton. Although most of the company's products were still in great demand the market for bellows and bellows operated forges had declined to the point where their production was no longer viable. In 1960 this department, which employed just 5 men, two of whom were then over 80 years of age, was closed down. Both of these men had worked for the company all their working lives, one having started with William Allday & Sons and the other with J.C.Onions. The Foreman of this department, Harry Hopkins, was also retired at this time, having been employed by the company since the age of 13 when he had worked under his father who was also an old J.C.Onions man.

The Clean Air Act of 1956 was probably the first of a wide range of government measures to reduce the amount of industrial pollution being emitted into the atmosphere of our large cities. This affected Alldays in two ways. By 1968 the demand for 'in-house' castings had diminished, chiefly as a result of fan production using more fabricated construction, so this department too was closed. It was proving difficult to find skilled craftsmen for the foundry anyway, but the final deciding factor appears to have been the potentially high cost of re-equipping the department to meet the Clean Air Act and local bye-laws.

However, equipment required for installing in workshops to meet the conditions of the Clean Air Act had to be made by someone, so Alldays introduced some of this equipment into their own product range. The first product appears to have been a Scroll Collector to remove grit and dust from the exhausts of coal/coke fired steam boilers and furnaces. This was made up in three parts, a primary separator, a secondary collector and a dust storage and removal

DIAGRAM

1950 BLOCK PLAN OF THE "GREAT WESTERN WORKS"

A. Erected in 1940 this single storey building measured 62 feet by 18 feet, was 12 feet high to the eaves and 18 feet 4 inches high to the ridge. The architect was Ewan Harper and the steelwork for this building was made by Charles Wade Ltd. Although no designation was shown for this building a later drawing indicates that it was used as a Sawmill.

B. In 1948 a new Machine Shop was erected which required the partial demolition of a number of existing buildings. Built in two bays, 40 feet wide and 58 feet wide, the building was 112 feet long and featured two concrete ramps which measured 20 feet by 12 feet to cater for differences in site levels within the building. The architects were Ewan Harper & Co.

C. In 1950 the old Plate Shop was rebuilt to match building 'B' and was redesignated as a Machine Shop. The increased height of this new building necessitated blanking off the first floor windows in the 3 storey high building behind it.

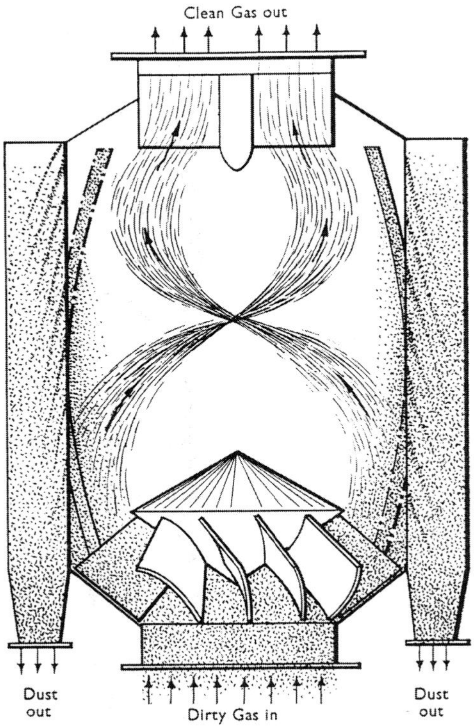

UF3 Dust Collector

View of vanes on UF3

The OMEGA Dust Collector Fan

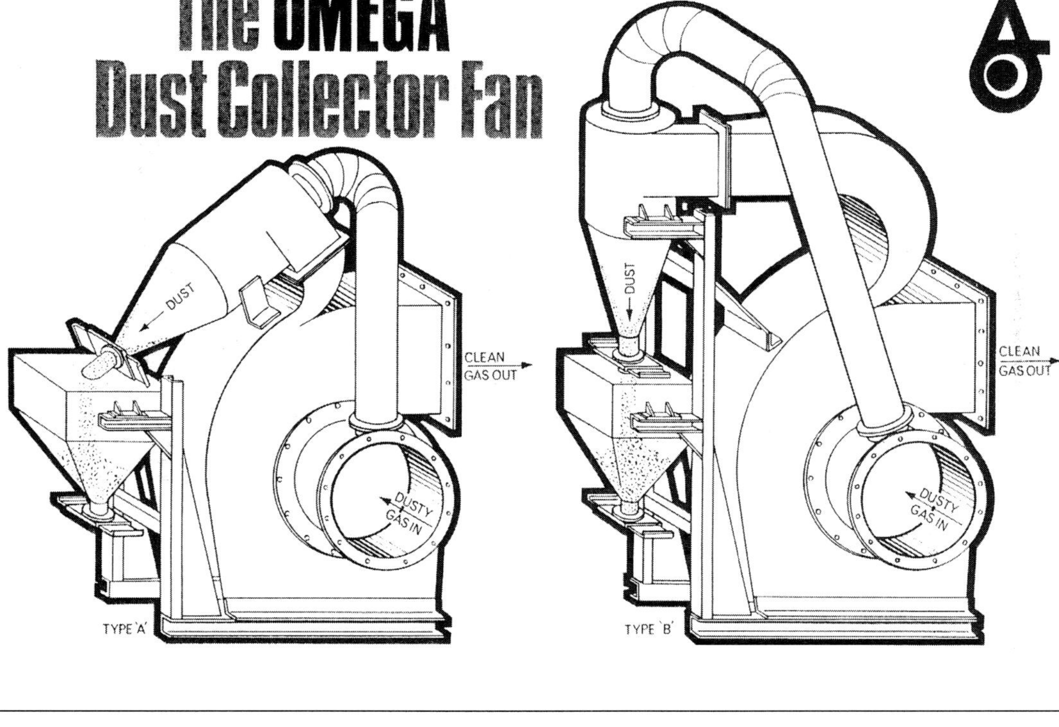

ILLUSTRATIONS

LEFT-TOP: A sectioned view of the UF3 dust collector showing how dirty gas entered the apparatus with a swirl action imparted on the gas by the vanes forcing it onto the walls of the collector and falling into the two ducts which then directed the dust to be collected and disposed of from a convenient point on the ground.

LEFT-BOTTOM: General views of the "Omega" dust collector, a more compact system for dust collection which could be floor mounted as a complete unit.

PHOTOGRAPH

BELOW: A general view of part of the machine shop where various castings and other parts were produced covering the whole range of company products. Some of the castings in this picture were for some of the last NSK hammers to be made, parts of which were finished by draw-filing and hand scraping.

hopper. The primary separator received the whole output of dust laden gas, separated it into clean gas which it exhausted into the atmosphere, and concentrated dust laden gas which was transferred into the cyclone type secondary collector which removed all the dust which was deposited in a removable hopper and transferred the remaining clean gas into the exhaust section of the primary separator. The entire piece of equipment was designed to take up very little space, be very efficient in operation and be completely reliable.

The second piece of equipment, introduced about 1964, was the (Upward Flow) Dust Collector which was designed to fit into an existing chimney. Using their fan experience they designed a simple unit featuring high efficiency vanes which imparted a swirl to the dust laden gas forcing it onto the walls of the collector and at the same time separating the dust from the gas into two integral side chambers which were connected by suitable ducting to allow the dust to be collected and removed at a convenient point on the ground. For more difficult installations a horizontal version was available, the HF3, which worked on similar principles.

More compact, floor mounted, dust collectors were developed known as the 'Omega' Dust Collector Fan which was available in a range of seven different models to cope with gas volumes of 9,500 cubic feet per minute up to 50,000 cubic feet per minute and designed for a gas temperature of 440°F.

About this time too a special range of engineering equipment was introduced designed specially for technical schools and colleges where engineering instruction was still very much a 'hands-on' experience. The equipment included a number of different, small, forging and brazing hearths which could be supplied complete with

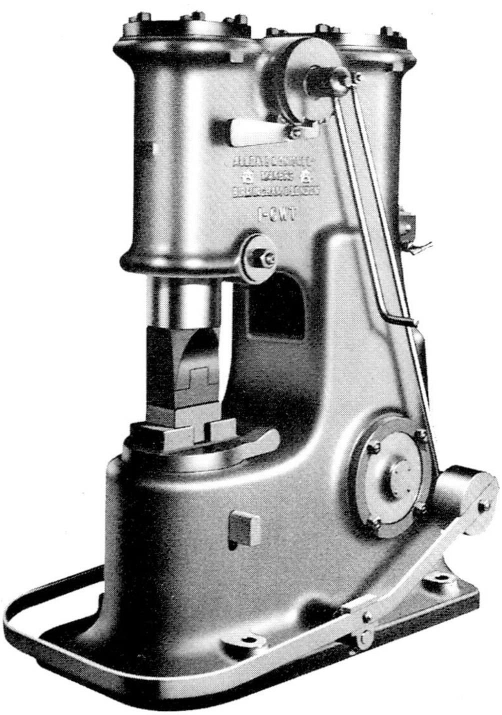
1-CWT

PHOTOGRAPHS

LEFT-TOP: A typical fan assembly fitted with Inlet Vane Control (IVC) on the input side of the fan. By closing the inlet vane control, using the hand operated lever clearly shown here, the pressure/volume characteristics of the fan could not only be controlled, but the running efficiency of the fan working on part loading could be improved quite dramatically. In short, this device could reduce the running costs of a typical fan installation when not required to work at its designed maximum output.

LEFT-BOTTOM: The company produced a wide range of engineering equipment designed specifically for workshops of technical schools and colleges. Shown here are just two examples of such equipment, a 'PB' 1/2 cwt pneumatic forging hammer with hand or foot operation and capable of delivering up to 165 blows per minute and, on the right, a 'MO105' model crucible tilting furnace for melting up to 60 lbs of non-ferrous metals. Using a gas or oil burner the furnace was fitted with an electrically driven blast type fan. As well as dwarf cupolas, forging hearths, bending slabs and a variety of blacksmiths' tools, a special fan assembly was offered complete with measuring equipment to allow laboratory testing and measuring of the characteristics of the fan to be demonstrated and measured.

grit arresters in their hoods, a revolving brazing hearth, a dwarf type cupola and a small crucible tilting furnace. Ancillary equipment for the college workshops also included a 1 cwt anvil, swage blocks, blacksmith's leg vice, hammering plates and bending slabs mounted on steel stands, sets of blacksmith's tools and small marking off tables. Where heavier work was being taught a very compact 'P.B.' 1/2 cwt pneumatic forging hammer was available driven by a 3 phase electric motor. Another piece of more specialised equipment was a laboratory air measuring unit intended for students entering the heating and ventilating industry, which enabled the performance characteristics of a centrifugal fan to be studied and demonstrated.

A more unusual product was introduced known as the "Jetstream Airveyor" (Regd) which was made under licence from an unknown company. This equipment consisted of box section ducts, with louvred tops, connected to a suitable fan, which conveyed light materials from one production point to another on a cushion of air. The ducts could be made in any shape or length enabling material to be conveyed between different departments or buildings. Usually only a single fan would be required to provide the cushion of air, but where more lengthy conveyors were required extra booster fans could be introduced at strategic points. Applications for this novel equipment included the removal of waste in the cardboard box industry, conveying packets or boxes, conveying plastic tiles and conveying cereals. The conveyed materials could also be cooled or dried by the air during transit through the Airveyor system or, with the addition of a heated plenum between the fan and conveyor, goods or materials could be heated.

SWL 10 CWT No 21

FAN DESIGN & DEVELOPMENT

The pre-war range of 'LS' high pressure fans was developed into a much larger range of sizes, from a size 12 up to size 40, with much more emphasis on fabricated casings rather than the earlier cast iron casings, a trend which eventually led to the total abandonment of the cast iron designs.

The 'LS' range started at a 2/14 model which produced 20 cubic feet of air per minute up to the 12/40 model which produced a maximum of 10,000 cubic feet of air per minute, all the 'LS' models being driven at 2,900 r.p.m. This range of fans could be fitted with the special A&O bearing gas head to prevent any possible leakage of poisonous or explosive gases and was also available as a 2 stage fan for applications requiring extra high pressures.

The 'GP' range of fans was designed for general purpose industrial applications and could handle dusty air, gases, corrosive or abrasive gases, and could be used for suction or pressure applications, indoors or outdoors. The fabricated casings and impellers could be supplied with rubber or plastic coatings when required for working corrosive or abrasive gas. This range comprised six basic models which started at the GP5 model which could produce 423 cu.ft./min. up to 4,225 cu.ft./min., to the largest GP10 model which had a range of 1,521 cu.ft./min. up to 17,494 cu.ft./min. The drive to these fans could either be directly by electric motor, or by vee belts driven by an electric motor, according to the fan size and type of application.

About this time the company developed a control system for their fans known as 'IVC', or Inlet Vane Control, which could produce very substantial savings in fan operating costs where the fan was not required to provide its maximum designed output 100% of the time. When a fan was not required to produce its maximum output all the time the normal practice was to fit the fan with a crude damper control, either on the inlet or discharge side, the damper being, more or less, a throttle control to prevent the fan delivering its maximum output. With the electric motor still driving the fan at its maximum speed the whole machine was then being operated very inefficiently. By introducing the specially designed inlet vane control (IVC) to the input side of the fan and specifying a 2 speed electric motor the air flow characteristics into the fan itself could be changed allowing the fan to be operated at savings estimated at up to 44%. The inlet vane control unit itself comprised a circular assembly of specially designed vanes, or blades, operating rather like the iris in a camera to provide an infinitely variable opening to the inlet side of the fan. This device could be operated manually or hydraulically, according to the size of the fan, either at the fan itself or remotely according to the position of its installation, and could also be designed as a completely automatic control system which operated the inlet vane control and the electric motor driving the fan.

By 1966 the "Powervane" range of centrifugal fans had been introduced to provide large volumetric output in restricted installation spaces. This range was suitable for industrial ventilation, providing draught for small boilers, exhaust or pressure units for dust collection plants and combustion air or recirculation for large furnaces. All models were designed to be driven by electric motors which could be of direct or indirect drive, with the motor horizontal

PHOTOGRAPHS

Both of these photographs were taken at the Sydenham Road works in the mid 1960s.

The top picture is almost an aerial view of the Machine Shop and shows two of the enormous lathes to the left, still belt driven from an overhead lineshaft, but, by this time of course, the lineshaft was driven by electric motor. To the right of this picture is one of the horizontal boring machines which would have been used for a multitude of jobs.

The lower picture shows the Impeller Assembly Shop with a wide variety of impellers lined up across the shop floor and in the centre of the picture a radial drilling machine had been set up to work on the sheet of steel supported by a wooden crate.

In the lower right hand corner, behind the smith's hearth, are sheet steel fan casing parts which had been cut to shape on a profile cutting machine. The section of this building to the right with the much lower roof line was built in 1929 as a Light Machine Shop, and the remainder of the building was built in 1938 as a Heavy Machine Shop.

or vertical and for special high temperature applications air cooled or water cooled bearings were available. The fan casings were constructed of welded heavy steel plate adequately stiffened to prevent drumming with the scroll specially designed to reduce turbulance and with the inlet cone designed to give an even flow of gas or air into the impeller. The standard "Powervane" models ranged from a number 5 through to a size 19 with outputs of 1407 cu.ft./min. up to 135,000 cu.ft./min., which required fan horse powers of 0.178 to 295. Four larger models, numbers 20 to 23, were made to special order.

Fans designed for medium to heavy duty applications with large volumes of air or gas at low pressure were introduced known as the 'AA' type which were suitable for operation at up to 500°F in standard form. Like most Alldays fans at this time the 'AA' type could be specified with belt drive and direct or indirect electric motor drive and with double inlet, double width. Standard models produced outputs of 3,000 cu.ft./min. up to 178,400 cu.ft./min., sizes 14 to 26, with larger models made to order. The largest single inlet single width fan measured 11ft 5ins long by 13ft 6ins high by 8ft wide and the largest double inlet double width model measured 11ft 11ins long by 15ft 1in high by 11ft 4ins wide.

For the heating and ventilating industry a completely new range of fans was developed, the 'LV' type (Low Velocity), in a range to cover the smallest to largest installation, from 103 cu.ft./min up to 293,333 cu.ft./min. However, for even larger installations a double inlet double width(D.I.D.W.) 'LV' type could be specified with an output of 500,000 cu.ft./min. This fan had a 10 feet diameter impeller, measured 14ft 5ins long by 17ft 5ins high by 19ft 2 ins wide, weighed 12 tons 11 cwts and required 692 fan horse power. Construction of the casings, particularly the larger models, featured components bolted together which could be dismantled to allow fans to be built up piece by piece in particularly restricted installations. 'LV' fans were used in such buildings as The New Bank of England, London, Rackhams New Store, Birmingham, Lewis's Ltd., Manchester and even Mitchell Cotts own headquarters in London. Typical prices for 'LV' fans in 1970 were £83 for a number 13½ with inlet damper and powered by a 2 h.p. electric motor running at 950 r.p.m., £179 for a number 18 powered by a 3 h.p. motor giving an output of 5,000 cu.ft./min at 775 r.p.m. and with the belt driven fan and motor mounted on a single bedplate. A number 30 fan with belt drive from a 25 h.p. electric motor mounted on a bedplate complete with anti-vibration pads and with an output of 18,000 cu.ft./min at 690 r.p.m cost £330-10s-0d.

Another range of Alldays designed fans still in production at this time was the 'CRB' type which were made in a large number of different sizes, such as a number 135 powered by a 1 h.p. electric motor driving the fan at 2,900 r.p.m. and with an output of 450 cu.ft./min. at £62, a number 156 powered by a 3 h.p. electric motor with an output of 1,000 cu.ft./min at £89 and a number 183 gas booster powered by a 1 h.p. motor driving the fan at 2,900 r.p.m. and giving an output of 160 cu.ft./min at 12" boost pressure at £130.

Many of these smaller types of fans were ordered directly by contractors carrying out engineering installations at all kinds of factories, like sawmills, cement works, glass works and iron works, in Britain and abroad, like Chile, Africa and Korea.

PHOTOGRAPHS

TOP: A Double Inlet Double Width fan (D.I.D.W.) with the top half of the casing removed to show the impressive impeller assembly. The impeller is made up of a large number of shallow blades which are curved forward in the direction of rotation and supported by a central drive boss incorporating six angle section arms. The whole impeller and drive shaft assembly is supported by two substantial bearings mounted on blocks, there being no physical contact between the impeller and fan casing, to reduce vibration and noise. The fabricated casing is bolted together making erection and servicing of the fan much easier.

BOTTOM: A special purpose, one-off fan supplied complete with drive motor and fabricated bed plate to customers' own specification.

A
A
H1872
OK
C

PHOTOGRAPHS

LEFT-TOP: Taken in 1965 this photograph shows Keith Thomas working on a size 18 B.A.M. fan made for Simon-Carves, The British Sugar Corporation at Bardney, Lincolnshire. The fan was driven at 940 r.p.m. by a 185 h.p. electric motor via a Sinclair hydraulic coupling, absorbed 166 fan h.p. and had an output of 78,000 cu.ft./min.

Standing in the foreground are 4 of a batch of 18 44"diameter bifurcating axial flow fans made for Colvilles Gartosh Works for supplying cooling air to steel plate coils undergoing annealing in vertical heat treatment furnaces. These fans were designed to run at 960 r.p.m. with an output of 20,000 cu.ft./min.

LEFT-BOTTOM: Fitter Norman Stone is seen standing near to some fan casings and an impeller giving some idea of the size of the enormous fans being made in the works during the 1960s. The lever assembly attached to the upper fan casing would eventually be coupled up to the five shafts protruding from the casing which operated dampers.

ILLUSTRATIONS

RIGHT: Some of the drive arrangements for the "Powervane" range of fans. The top illustration shows a fan designed to recirculate air inside a furnace and very often, because of the very high operating temperatures, it was necessary to provide water cooled bearings and drive shaft.

The centre illustration shows a more conventional arrangement with the fan driven by an electric motor via a flexible coupling and with the fan impeller supported on the two bearings mounted on top of the fabricated motor stool. This particular fan is fitted with an inlet vane control (I.V.C.).

The bottom illustration shows a fan assembly where the impeller shaft is supported by bearings to the left and right of the fan casing itself, with again an electric motor driving the fan via a suitable coupling.

Gas fans could also be supplied where the electric motor was positioned in a chamber separate from the fan assembly itself, for reasons of safety, and the drive shaft passed through a wall with suitable seals to prevent any risk of fire or explosions.

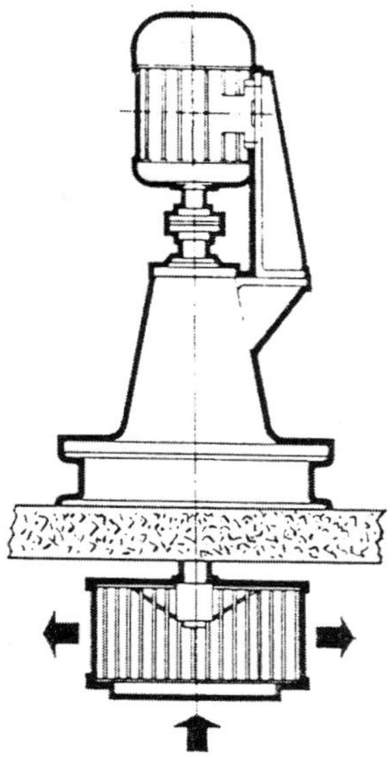

Recirculating fan inside furnace

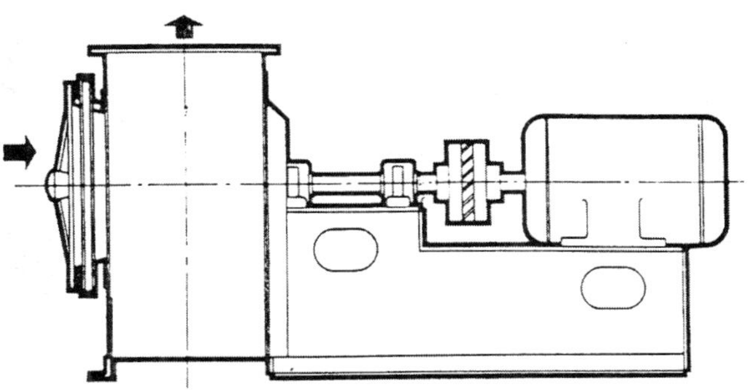

I.V.C. controlled fan, supported on casing feet and base

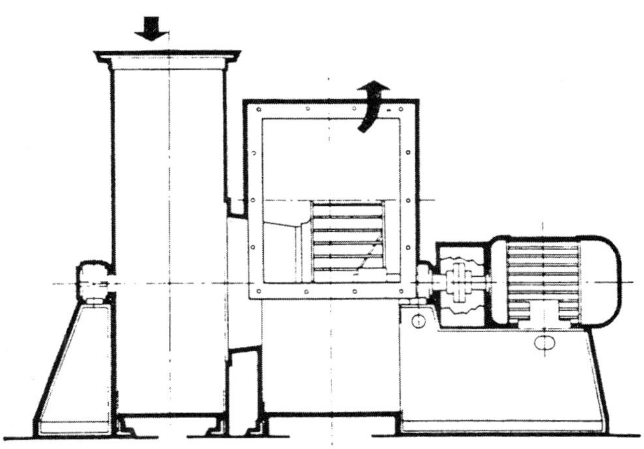

Box inlet fan, outboard bearing independently supported

ALLDAYS PEACOCK & CO.LTD.
Recent History

In 1946 one of the Alldays & Onions senior fan designers, Bill Morris, left the firm and together with John Terry, who had a business agency, set up a new company J .C .Peacock (Engineers) Ltd, to make industrial fans in premises quite close to Birmingham city centre, in Clement Street. As is usual in these instances a number of Alldays' employees departed with Bill Morris to join the new company. These included Les Dealey a sheet metal worker, Graham Hill a toolroom worker and Mr Buckley from the machine shop.

Making a similar range of fans to Alldays this new company obviously prospered in the post-war boom years and in 1953 a new factory was bought in Ballet Street, Smethwick and the name of the company was soon changed to J.C. Peacock (Aerodynamic Engineers) Ltd. In 1962 the company relocated to a large industrial estate in Weston-super-Mare where it specialised in production of 'QT' and 'LS' type high pressure fans. The move to Weston had been assisted in no small way by the provision of local council incentives which offered council houses to relocated employees and attractive grants to companies for purpose built factories. There were also government incentives at this time for employing apprentices in industry, and one 16 year old young man to join the company as an apprentice, in 1962, was Bob Hockey, who worked his way up through the company to become its Managing Director in 1989. Quite a number of J.C. Peacock's highly skilled employees, like Gordon Burns a sheet metal worker, had moved to Weston with the company and it was through these men that skills were passed on to apprentices who spent 6 months in each department, such as machine shop, tool room, impeller section, fan fabrication, fitting shop and drawing office. Bob's fellow apprentices included Martin Blackmore, who is now Works Manager, Clem Hopkins, and Brig Leslie, who is now Machine Shop Supervisor.

Mitchell Cotts & Co.Ltd., the holding company of Alldays & Onions Ltd., acquired the J.C.Peacock company early in 1969 and, by the end of the same year, had united it with Birmingham based Alldays & Onions Ltd., to form a new company, Alldays Peacock & Co.Ltd., specialising in all forms of industrial fans. Naturally, there was some duplication of products in the ranges of the two companies and a rapid rationalisation programme was instigated which saw volume production of high pressure combustion fans, and backward inclined fans up to 1000mm in diameter, transferred to Weston with the Birmingham factory in Sydenham Road left to concentrate on larger, specialised, fans in its huge workshops.

As a result of the coal miners' strike in 1972 many companies found themselves working a 2 day or 3 day week in an effort to conserve power, but the Sydenham Road factory worked as normal due to a special dispensation granted to them whilst building five special fans for the United Kingdom Atomic Energy Authority. These 2 stage fans had casings made from 25mm thick steel plate, were equipped with special seals to prevent gas leakages and had an output of 7,500 cu.ft./min. There was an abundance of work from U.K. traditional customers like I.C.I. and British Steel, particularly the latter whose orders took up 2 months of the company's production each year. However,with the slow decline in the British steel indus-

PHOTOGRAPHS

TOP: A general view of the huge machine shop and fan assembly shop taken in the 1970s at the Sydenham Road works. In the right foreground a milling machine is being set up to machine a motor stool for the fan casing which is also mounted on the machine. To the left is a large lathe used to turn the fan impeller shafts, some of which can be seen on the machine shop floor, with George Stokes, the Machine Shop Foreman, standing by.

BOTTOM: One of the five 2-stage high pressure fans made in 1972 for the United Kingdom Atomic Energy Authority. The casings were made from 25mm thick steel plate and had an output of 7,500 cu.ft./min. at 95" SWG, and were used for circulating argon coolant gas, in an environment requiring such fans to be guaranteed 100% leak proof.

ALLDAYS PEACOCK

PHOTOGRAPHS

LEFT-TOP: Stan Donnington the Fan Testing Department Supervisor takes a final look over two inert gas fans made for crude oil tankers. These particular fans were driven by steam turbines, shown here coupled up to the fans, which were bought in from a proprietary manufacturer.

LEFT-BOTTOM: Fitter Michael Fitzpatrick stands alongside a huge "Europa" type fan being prepared for testing prior to being despatched to the customer. Note the ten belts used as the drive between the electric motor and the pulley on the fan impeller shaft!

With metrication being adopted in the company, all data was being quoted in metric, so this range of fans had quoted outputs up to 510,000m/hour at a pressure up to 889mm w.g. Hot gas versions of this fan were available to handle gas temperatures of up to 750°C.

BELOW: The 'V' range of axial flow fans was made under licence from DLK of West Germany. This range of fans was designed for medium and high pressure with the 'VM' type producing up to 160,000m/hour at 120mm w.g. and the 'VH' type fans producing up to 90,000m/hour at 700mm w.g.

try, the company had to explore other outlets for its products and diversified into building fans for industries such as flour milling, conveying fans, and high temperature recirculating fans for kilns and pottery applications.

The Birmingham factory soon found another new application for its centrifugal fans in the oil industry where the latest supertankers used for carrying crude oil required a new method of fire and explosion protection using inert gas. The lethal mixture of air and vapour rising from the crude oil presented serious fire hazards, but an inert gas of nitrogen/carbon dioxide, which was normally exhausted from the funnels of steam driven tankers as a combustion by-product, was capable of being tapped off, cleaned and then delivered under pressure to the void above the crude oil level in the storage tanks, so preventing the danger of fire or explosion. Specially designed Alldays fans were built to extract the inert gas from the 'Scrubber', or gas cleaner, and deliver it to the storage tanks under a pressure of about 2,000mm w.g. The fans had to be capable of withstanding internal erosion by the salt water laden gas from the 'Scrubber', be absolutely sealed against any possible leakage of the inert gas into other parts of the vessel and be capable of working in every kind of weather condition experienced by these tankers during voyages. These stringent operating conditions meant that the fan casings had to be coated internally with epoxy or rubber and the impellers needed to be made from nickel aluminium bronze, stainless steel or titanium.

Marine electric supply at 60Hz, as opposed to domestic 50hz supply, also raised problems for the company's test engineers, as

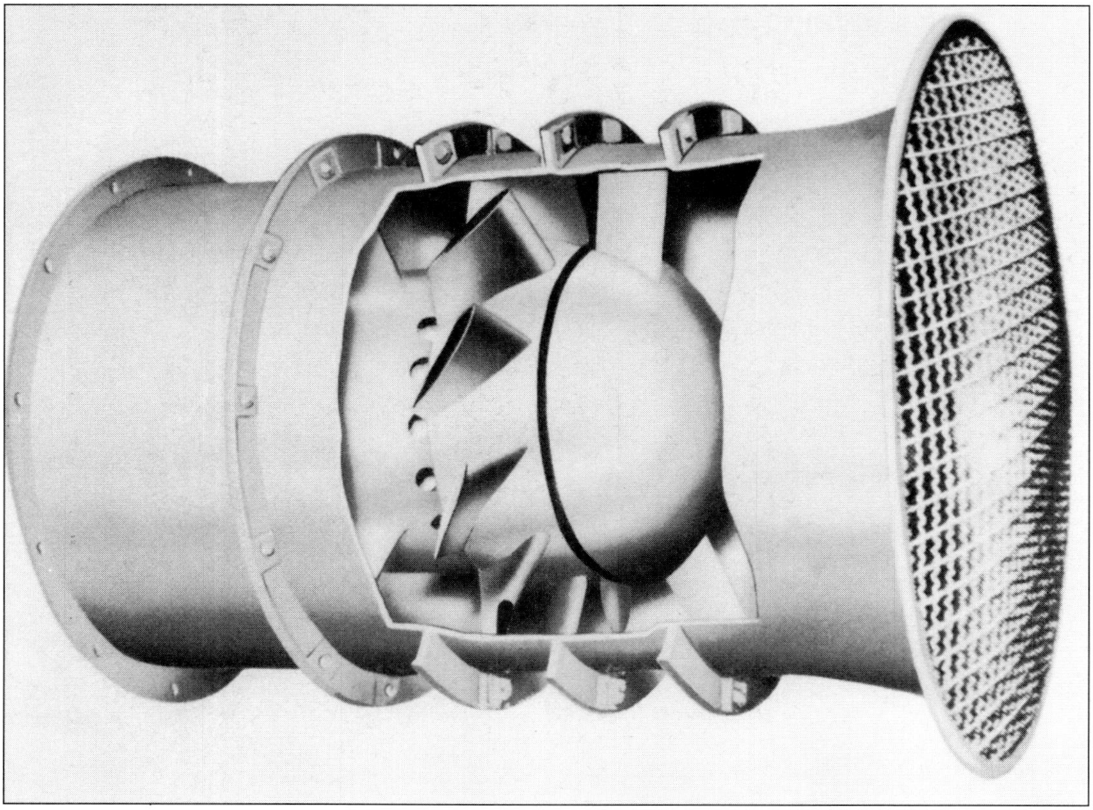

every fan had to be fully tested before leaving the factory. A purpose built generator set was bought powered by a triangulated 18 cylinder diesel engine rated at 2,500 h.p. and installed in a small building built alongside the canal side of the three storey workshop to give a 60 Hz supply to the special test cell built inside the ground floor of this building and capable of testing fans to a 100% over-run condition ensuring that the product would be of the highest quality and reliability when in operational service. With the closure of the old foundry in 1968 the buildings had remained unused and so, in 1972, the decision was made to demolish this historic part of the Great Western Works and use the site as a car park. However, in other respects the company had to move with the times, and an energetic ongoing research and development programme has kept the company to the forefront of fan design technology, especially with the manufacture of fan impellers with greater efficiency. With radiated noise becoming an increasingly important issue Alldays Peacock developed a complete range of silencers and acoustic ancillaries to meet the most stringent of regulations.

About 1974 the company supplied two fans to Mackley Ace of Southampton for an air cushion transporter being built for use in the Gulf where it was to ferry liquid natural gas plant between Abu Dhabi and Das Island. The craft, which had to be towed by tugs, was 180 feet long by 80 feet wide, weighed 750 tons, carried a 250 ton payload and could be towed at speeds up to seven knots. The two 1400 B.A. fans delivered 135,000 cu.ft. of air per minute at 32" static w.g. pressure and were driven by two 900 b.h.p. V-12 diesel engines.

PHOTOGRAPHS

LEFT-TOP: A general view of the fan casing fabrication shop, together with the sheet metal machine shop in the far background.

LEFT-BOTTOM: An excellent aerial view of the fan assembly shop with a variety of different fan types completed and ready for testing.

BELOW: The 'Deltic' generator set being unloaded at the Great Western Works where it was installed to enable on-site testing of marine inert fans before they were despatched for installation in oil tankers. The exhaust system and silencers for this special generator set which were removed for transportation can be seen on the ground.

406

At the Weston factory a section of the works had been given over to the refurbishment of gearboxes which increasingly required more space, necessitating the building of a new bay to house specialised equipment and test areas. After 3-4 years the company moved to another site leaving gearbox refurbishment completely autonomous from Alldays, and the parent holding company renamed this side of the business Mitchell Cotts Precision Engineers.

A prolonged recession in the early 1980s plunged the company into a difficult financial position and forced it to shed many of the 300 jobs then shared between the Birmingham and Weston sites. In January 1986 Mitchell Cotts sold the Alldays Peacock company to the Midlands based Spire Group headed by their Chairman Mr Gordon Brown who was a very experienced accountant.

By 1988 Gordon Brown had transferred all production to the Weston-super-Mare site and sold off the Sydenham Road premises. A great challenge now presented itself to the staff in Weston to prove that they could design and manufacture fans of all types, and to the

PHOTOGRAPHS

LEFT-TOP: A piece of history disappears as the old foundry buildings at the Great Western Works are torn down in 1972 to provide space for a car park.

LEFT-BOTTOM: Mr James Tye (right) Director General of the British Safety Council is seen here presenting a Safety Award in 1978 to Peter Thomas (centre) Works Engineer for the Sydenham Road works and Broderick Baughan, also from the Sydenham Road works.

RIGHT: This picture shows a typical installation of an Alldays Peacock fan, shown here with the guard removed from the multi-belt drive during maintenance work in 1966 at the Lowestoft premises of Bolton Paul Ltd.

highest quality standards, in the tradition of the old Alldays company. The highly skilled workforce soon exhibited their mettle and orders improved to the point where an additional bay was added to the works and extra people were recruited, some of whom had previously worked for Alldays. A 10 tonne overhead crane was installed, which, together with the acquisition of a CNC punch nibbling machine, increasingly improved productivity and profits, and with the West Bromwich sales and drawing office staff transferred to the Weston site further improvements were accomplished in efficiency and planning. Ray Stallard, the Sales Director, re-joined the company, having previously worked at Sydenham Road, and new U.K. sales managers and agents abroad ensured an ever increasing order book, and the company once again started attending commercial exhibitions to regenerate business, sometimes from old customers who had thought the company was no longer operating.

An enquiry from the company's agent Adiran, in Tel Aviv, Israel, resulted in the biggest single order obtained since being in the ownership of the Spire Group. Eight high pressure dilution air fans were installed in a new plant built by Dead Sea Magnesium, for refining magnesium extracted from the Dead Sea. A further order for three fans followed for the same project, together with the supply of acoustic equipment and spares, and members of the Alldays management were invited to attend the inaugural opening ceremonyin 1987. Mitsui Babcock placed two large orders for fans to be used in power stations being built in India and Vietnam, followed by orders for further power station projects in Dang Dong and Phai Lai in China.

The Weston-super-Mare factory continues to manufacture fans to very high quality standards suitable for a multitude of industrial and marine applications. Their range includes the type 'BL' and 'BA' with impeller diameters from 400mm up to 2,000mm with outputs from 5,500 cult./min. up to 360,000 cu.ft./min. The 'QT' and 'CRB' range still includes models suitable as blowers or gas boosters and, as well as 'DVFC', 'DVRC', 'BC' and 'PB' ranges, the company is still able to offer more specialised ranges of fans suitable for gas coolant, inert gas and hot gas. Over the past 15 years the company has adapted to a changing business spectrum and globalising markets have led to different attitudes toward customer requirements, with internet sites and computerised fan selection programmes giving greater access to overseas customers and improvements in response time to enquiries.

The servicing of fans has become a larger part of the company's business, and the installation of on site dynamic balancing and vibration monitoring systems fitted to fans gives customers advance warning of impending problems which alleviates extended down time.

Ironically, the Great Western Works in Birmingham is now owned by a company called Wallwork Heat Treatment (Birmingham) Ltd., who specialise in heat treatment and use equipment which, had Alldays survived in its original form, would probably still be being made on this site!

The final word about this historic company comes from Managing Director Bob Hockey. "The company was founded on good working principles with good management and a skilled workforce, this has meant it has survived to this day, and if these fundamental principles are maintained I see no reason why it cannot continue for a further 350 years."

Diesel engine driven fans for hover platform. This platform was used for lifting aircraft, which had overshot the runway into a swamp, in Nigeria. There were eight fans for four platforms.

This fan has a 3.3.kV motor, the unit produces air at a very high pressure for a gas plant in Korea.

The impeller is manufactured from aluminium bronze. We now manufacture in titanium, Hasteloy, all types of stainless steel like 304, 321, 316, 316L, etc. Also incalloys, abrasive resistant steels as well as mild steel.

A 50mm thick fibreglass casing with a titanium impeller and shaft was needed to move air contaminated with noxious gas through a wet scrubbing system. The end product, after scrubbing, was clean enough to be expelled into the atmosphere. Special sealing around the shaft entry to the casing was required as the air mixture could kill.

Wherever toughened glass is supplied it has passed through a process of heating and fast cooling, this cooling is produced by a fan of this type. Usually used with a D.C. motor and controller, this fan will run variable speeds to suit the glass requirements.

This specialised two-stage fan was designed to fit into a tight space. This unit produced high air pressure for a foundry cupola system.

Fan on test at Alldays works. Due to fan handing the roof panels had to be removed to allow the correct length of test ducting to be fitted. The test engineers had to stand on ladders on the roof to reach the test points in the duct. The unit was for Yugoslavia.

These colourful fan units were for a magnesium plant in the Dead Sea works in Israel. They were fitted with 750 kW motors and had damper filter/silencers and inlet vane controls. The wife of the plant owner did the colour co-ordination.

Double inlet double width fan for British Sugar, Peterborough. This was one of the main induced draught fans for the sugar refining plant.

A set of fans for a gluten dryer system in Holland. Fans are finished to food standard.

Fans can be supplied with the case coated in plastic phenolic resins or as the fan indicated above rubber. This is to stop corrosion of the mild steel case, a stainless steel impeller is supplied to withstand attack by the acids in the airstream.

All fans were tested mechanically even though the design made things awkward for setting up. This is carried on today at our new premises in Weston-super-Mare using a more modern method inverter.

This fan on test at our works was destined for Japan on a rock crushing extract plant. The impeller was manufactured in abrasive resistant material and the fan is the largest belt driven unit manufactured by Alldays ever, the motor being a *550* kW.

This fan manufactured in the late 1950s is still working today. It has an inlet vane control fitted to the inlet, which was one of the first energy saving devices for air control.

ALLDAYS/ENFIELD CAR TYPE NUMBERING

From what remains of the company records, and from the few motor car catalogues accessed whilst carrying out research into the company's motor car production, the following information gives details of motor car type numbering from 1905 to 1913.

NUMBER 1.
1905 - 7 h.p. 2 seater on short chassis.
1905 - 8 h.p. 2 seater on long chassis.

NUMBER 2.
1905 - 7 h.p. 2 seater on short chassis.
1905 - 8 h.p. 2 seater on long chassis.
1908 - 10 h.p. 2 seater on short chassis.
1910 - 10/12 h.p. 2 seater on short chassis.

NUMBER 3.
1905 - 7 h.p. 1-3 seater on short chassis.
1905 - 8 h.p. 1-3 seater on long chassis.

NUMBER 4.
1905 - 8 h.p. 3-4 seater, swing seat on long chassis.
1905 - 10 h.p. 3-4 seater, swing seat on short chassis.
1910 - 10/12 h.p. 4 seater, swing seat on short chassis.

NUMBER 5.
1905 - 8 h.p. 2 seater with folding hood on long chassis.

NUMBER 6.
1905 - 8 h.p. 4 seater on long chassis capable of being converted to 5 cwts van.

NUMBER 7.
1905 - 8 h.p. 4 seater with Cape Hood on long chassis.

NUMBER 8.
1905 - 8 h.p. 4 seater with detachable canopy for hot climates, on long chassis and available with paraffin carburettor.

NUMBER 9.
1905 - 8 h.p. open 4 seater on long chassis for hot climates and available with paraffin carburettor.

NUMBER 11.
1910 - 14/18 h.p. 2 seater with hood for Commercial Travellers.

NUMBER 12.
1908 - 10 h.p. 2 seat Commercial Traveller.
1910 - 10/12 h.p. 2 seat Commercial Traveller.

NUMBER 13.
1910 - 10/12 h.p. 2 seater with box body on back fitted with drawers for Commercial Traveller.

NUMBER 14.
1907 - 10 h.p. Roi de Belge.

NUMBER 15.
1905 - 10 h.p. 2/4 seater, open car, with detachable Tonneau.

NUMBER 16.
1908 - 10 h.p. 2 seater van for 7 cwts or 10 cwts.
1910 - 10/12 h.p. 10 cwts van.

NUMBER 17.
1910 - 14/18 h.p. 15 cwts van.

NUMBER 18.
1908 - 14 h.p. 4 seater open car on 8'-6" wheelbase chassis.
1910 - 14/18 h.p. 4 seater with hood Semi-Torpedo body.
1912 - 16/20 h.p. 5 seat Semi-Torpedo on 9'-0" wheelbase chassis.

NUMBER 19.
1910 - 14/18 h.p. 2 seater "Brooklands".

NUMBER 20.
1908 - 20 h.p. 4 seater with hood, Roi de Belge on 9'-0" wheelbase chassis.

NUMBER 21.
1907 - 10 h.p.

NUMBER 30.
1908 - 10 h.p. 4 seat Landaulette l.w.b.

NUMBER 31.
1908 - 14 h.p. 4 seat Landaulette.
1910 - 14/18 h.p. 4 seat Single Landaulette.
1912 - 16/20 h.p. Single Landaulette.

NUMBER 32.
1908 - 14 h.p.

NUMBER 33.
1908 - 20 h.p. 4 seat Landaulette 9'-0" w.b.
1910 - 20/25 h.p. Single Landaulette.

NUMBER 34.
1908 - 10 h.p. 4 seat Landaulette, open cab.

NUMBER 35.
1908 - 14 h.p. 4 seat Landaulette, open cab.
1910 - 14/18 h.p. 2 seat Taxicab.

NUMBER 36.
1908 - 20 h.p. 4 seat Landaulette, open cab.
1910 - 20/25 h.p. Double Landaulette.

NUMBER 37.
1910 - 20/25 h.p. Limousine.

NUMBER 40.
1908 - 10 h.p. 4 seat Colonial Car.
1910 - 10/12 h.p. 4 seat Colonial Car.

NUMBER 41.
1908 - 14 h.p. 4 seat Colonial Car.
1910 - 14/18 h.p. 4 seat Colonial Car.

NUMBER 42.
1908 - 20 h.p. 4 seat Colonial Car.
1910 - 20/25 h.p. 4 seat Colonial Car.

NUMBER 45.
1908 - 20 h.p. 20/30 cwts van.
1910 - 20/25 h.p. 20 cwts van.

NUMBER 46.
1910 - 20/25 h.p. 30 cwts van.

NUMBER 55.
1908 - 10 h.p. 2 seat, closed, Doctors Coupé.
1910 - 10/12 h.p. 2 seat, closed Coupé.

NUMBER 56.
1908 - 14 h.p. 2 seat, closed, Doctors Coupé.
1910 - 14/18 h.p. 2 seat, closed, Coupé.

NUMBER 88.
1908 - 10 h.p. 4 seater with fixed canopy.

NUMBER 101 "ENFIELD".
1910 - 10/12 h.p. open 2 seater.

NUMBER 102 "ENFIELD".
1910 - 10/12 h.p. 2 seater with hood.

NUMBER 104 "ENFIELD".
1910 - 10/12 h.p. open 4 seater.

NUMBER 112.
1912 - 12/14 h.p. 2 seat "Brooklands" 9'-0".

NUMBER 116.
1912 - 16/20 h.p. 7 seat Torpedo, 9'-9" w.b.

NUMBER 123.
1912 - 12 h.p. 4 seat Torpedo, 9'-0" w.b.

NUMBER 125.
1912 - 25/30 h.p. 5 seat Torpedo, 9'-9" w.b.

NUMBER 161 "ENFIELD".
1910 - 16 h.p. open 4 seater.

NUMBER 181.
1910 - 14/18 h.p. 4 seat open Torpedo.
1911 - 14/18 h.p. Torpedo.
1912 - 16/20 h.p. 5 seat Torpedo.
1910 - 18/22 h.p. 4 seater "ENFIELD".

NUMBER 182 "ENFIELD".
1910 - 18/22 h.p. open 4 seater.

NUMBER 183 "ENFIELD".
1910 - 18/22 h.p. closed Landaulette.

NUMBER 186.
1912 - 16/20 h.p. Cabriolet Landaulette on 9'-9" w.b. chassis.

NUMBER 191.
1910 - 14/18 h.p. 2 seater with hood, Victoria or Doctors.
1912 - 16/20 h.p. 2 seat Victoria.

NUMBER 201.
1910 - 20/25 h.p. 4 seat Torpedo with hood.

NUMBER 212.
1912 - 12/14 h.p. 2 seat Victoria 9'-0" w.b.

NUMBER 221.
1908 - 10 h.p. 4 seater with hood, 8'-0" w.b.
1910 - 10/12 h.p. 4 seater, Semi-Torpedo.

NUMBER 222.
1908 - 10 h.p. open 2 seater, 7'-6" w.b.
1910 - 10/12 h.p. 2 seater with hood, Victoria or Doctors.

NUMBER 223.
1910 - 10/12 h.p. open 2 seater "Brooklands".
1912 - 10/12 h.p. open 2 seater "Brooklands".

NUMBER 312.
1912 - 12/14 h.p. 4 seat Torpedo, 9'-0" w.b.

NUMBER 412.
1912 - 12/14 h.p. 5 seat Torpedo, 9'-0" w.b.

NUMBER 525.
1911 - 20/25 h.p. 'D' front Landaulette, on a 10'-6" wheelbase chassis.

NUMBER 916.
1913 - 16/20 h.p. 4 seat Torpedo, 9'-9" w.b.

MOTOR CAR CHASSIS & ENGINE NUMBERING

Up to the introduction of the post war LC11 type, which had a new sequence of chassis numbers, motor car chassis numbering was on a consecutive sequence irrespective of type, horse power, or bodywork. Vehicles and engines, generally, appear to have been made in a small batch system, which would have been natural for a company building motor vehicles in, comparatively, small numbers, which would explain the use of suffixes to chassis numbers.

Engines and rear axles were also numbered consecutively and appear to have been used in any old order, as long as they suited the vehicle being assembled!

The information which follows is by no means a complete record of vehicle production in the time span covered, but is included for interest and general guidance. Some idea of the number of vehicles being produced can be judged by the chassis numbers recorded from 1748 in July 1907 to 4932 in April 1913.

SUFFIX	MODEL(S)	DATES DURING WHICH DESPATCHED
BP	10 h.p.	July 1907 to March 1908.
BS	10 h.p.	July 1907 to April 1908.
BS2	10 h.p.	March 1908 to July 1908.
BT	10 h.p. (For batch of chain drive chassis)	June and July 1907.
BU	8 h.p. & 10 h.p.	February 1908 to June 1908.
BW	20 h.p.	July 1907 to April 1908.
BY	10 h.p. & 14 h.p.	November 1907 to May 1909.
CD	20 h.p.	January 1908 to July 1910.
CF2	10 h.p. & 14 h.p.	December 1907 to December 1909.
CJ	20 h.p. chain drive charabancs.	March 1908.
CK	10 h.p. & 14 h.p.	July 1908 to May 1909.
EC3	12 h.p.	June 1911 to September 1911.
AA13	10 h.p.	August 1911 to March 1912.
AA14	10/12 h.p.	May 1912 to October 1913.
AB10	10 h.p.	July to September 1911.
AB11	10/12 h.p.	September 1911 to July 1912.
AB12	10/12 h.p. chassis.	April 1913.
AD9	20 h.p.	August 1911 to December 1911.
AD10	25/30 h.p.	December 1911 to December 1912.
ADV2	20 h.p. Charabanc & Van.	November and December 1911.
ADV3	20/25 h.p. Charabancs.	December 1911 to February 1912.
AC27	14 h.p.	June and July 1911.
AC28	14 h.p. & 16 h.p.	June 1911 to September 1911.
AC29	14/18 h.p.	July 1911 to January 1912.
AC30	14 h.p.	August 1911 to May 1912.
AC31	14 h.p.	September 1911 to November 1911.
AC32	14/18 h.p.	March 1911 to January 1912.
AC33	14/18 h.p. & 16/20 h.p.	November 1911 to March 1912.
AC34	14/18 h.p.	November 1911 to May 1912.
AC35	14/18 h.p.	January 1912 to March 1913.
AC36	14/18 h.p. & 16/20 h.p.	March 1912 to August 1912.
AE3	30/35 h.p. Bus & Charabanc.	November 1911 to August 1912.
AR1	10/12 h.p. Railcars.	October 1912 to April 1913.
B1	12/14 h.p.	November 1911 to February 1912.
B2	12/14 h.p.	January 1912 to February 1912.
B3	12/14 h.p.	February 1912 to April 1912.
B4	12/14 h.p.	March 1912 to May 1912.

B5	12/14 h.p.	February 1912 to September 1912.
B6	12/14 h.p.	April 1912 to August 1912.
B7	12/14 h.p.	May and June 1912.
B8	12/14 h.p.	June and July 1912.
B9	12/14 h.p.	June and July 1912.
B10	12/14 h.p.	July and August 1912.
B11	12/14 h.p.	August 1912 to May 1913.
B12	12/14 h.p.	September 1912 to June 1913.
B13	12/14 h.p.	September 1912 to December 1912.
B14	12/14 h.p.	November 1912 to July 1913.
B15	12/14 h.p.	December 1912 to July 1913.
B16	12/14 h.p.	October 1912 to April 1913.
B17	12/14 h.p.	January 1913 to July 1913.
B18	12/14 h.p.	March 1913 to August 1913.
CV1	14/18 h.p. Vans & 16/20 h.p. Bus.	February 1912 to September 1912.
CV2	14/18 h.p. Vans & Lorries,10/12 h.p. Van.	September 1912 to November 1912.
CV3	16/20 h.p. Vans & Buses.	October 1912 to December 1912.
D2	16/20 h.p. & 14/18 h.p.	January 1912 to June 1912.
D3	16/20 h.p.	January 1912 to August 1912.
D4	16/20 h.p.	July 1912 to September 1912.
D5	16/20 h.p.	September 1912 to December 1912.
D6	16/20 h.p.	October 1912 to December 1912.
D7	16/20 h.p.	November 1912 to February 1913.
D8	16/20 h.p.	December 1912 to August 1913.
D9	16/20 h.p.	January 1913 to July 1913.
D10	16/20 h.p.	February 1913 to July 1913.
E1	25/30 h.p.	August 1912 to February 1913.
EC4	12 h.p.	August 1911 to November 1911.
EV1	20/25 h.p. & 25/30 h.p. Charabancs.	March 1912 to May 1912.
EV2	25/30 h.p. Charabancs & Vans.	May 1912 to May 1913.
EV3	25/30 h.p. Buses & Vans.	November 1912 to November 1913.
RET	Electric Trolleybus Chassis.	June and July 1911.

ENGINE NUMBERS RELATED TO CHASSIS NUMBERS

The following information gives Engine Numbers quoted against the car Chassis Number into which it was fitted and also shows the Horse Power of the engine.

The first batch of numbers relates to some 12 h.p. motor cars which appear to have been built out of sequence and, interestingly, the engine numbers include odd and even numbers which are rarely found anywhere else in the remaining company records of motor car production.

Engine.	Chassis.	H.P.
744	3854	12
745	3856	12
746	3858	12
749	3862	12
750	3860	12
752	3864	12
789	3956	12
790	3868	12
809	3866	12
810	3870	12
811	3872	12
812	3958	12
813	3960	12
814	3968	12
815	3964	12
816	3966	12
817	3972	12
818	3970	12

Engine.	Chassis.	H.P.
2038	1824	10
2148	2022	8
2150	2016	8
2246	1806	10
2288	1758	10
2290	1756	10
2292	1794	10
2294	1918	10
2306	1854	10
2310	2034	10
2318	1780	10
2346	1852	10
2350	1848	10
2360	1752	10
2364	1904	10
2366	1800	10
2372	1856	10
2374	1748	10
2392	1858	10
2396	1762	10
2398	1750	10
2404	1766	10
2406	1760	10
2410	1776	10
2412	1930	10
2416	1898	10
2418	1770	10
2420	1764	10
2422	1754	10
2432	1772	10
2436	1782	10
2438	1778	10
2442	1768	10
2448	1774	10
2450	1814	10
2456	1788	10
2458	1850	10
2460	1786	10
2462	1784	10
2470	1792	10
2476	1790	10
2478	2026	10
2480	1802	10
2482	1804	10
2484	1820	10
2488	1936	10
2492	1812	10
2494	1796	10
2498	1816	10
2500	1810	10
2504	1798	10
2506	1868	10
2508	1882	10
2510	1866	10
2512	1938	10
2514	1872	10
2516	2094	10
2520	1920	10
2522	1826	10
2524	1880	10
2526	1878	10
2528	1818	10
2530	1808	10
2532	1822	10
2536	1870	10
2538	1828	10
2540	1896	10
2542	1874	10
2544	1894	10
2546	1890	10
2548	1876	10
2550	1836	10
2552	1884	10
2554	1888	10
2556	1838	10
2558	1892	10
2560	1830	10
2562	2090	10
2564	1914	10
2568	1832	10
2570	1934	10
2572	2096	10
2574	1906	10
2576	1912	10
2578	1916	10
2580	1922	10
2582	2092	10
2584	1910	10
2586	1908	10
2588	1900	10
2590	1902	10
2592	2072	20
2594	2070	20
2596	2076	20
2598	2074	20
2600	2066	20
2602	2068	20
2604	2088	20
2606	2080	20
2608	2084	20
2610	2078	20
2612	2082	20
2614	2086	20
2616	2102	20
2618	2104	20
2620	2106	20
2622	2120	20
2624	2112	20
2626	2196	20
2628	2108	20
2630	2118	20
2634	2142	20
2636	2126	20
2638	2116	20
2640	2198	20
2642	2114	20
2644	2128	20
2648	2144	20
2650	2148	20
2654	2130	20
2656	2132	20
2658	2124	20
2660	2122	20
2662	2134	20
2666	2138	20
2668	2146	20
2670	2136	20
2672	2152	20
2674	2164	20
2676	2156	20
2678	2160	20
2682	2174	20
2702	1932	10
2704	2190	10
2708	2180	14
2710	2268	14
2720	1834	10
2722	1926	10
2724	1842	10
2726	1924	10
2730	2018	10
2732	2024	10
2734	1840	10
2736	1846	10
2738	1844	10
2742	1952	10
2744	1948	10
2746	1946	10
2748	1942	10
2750	1950	10
2752	1944	10
2754	1958	10
2756	1956	10
2758	1954	10
2760	1966	10
2762	1968	10
2764	1962	10
2766	1960	10
2770	1940	10
2772	1972	10
2774	1964	10
2776	1980	10
2780	1974	10
2782	2212	10
2784	2038	10
2792	2204	10
2794	2028	10
2798	1978	10
2802	2042	10
2806	2032	10
2808	1984	10
2810	2020	10
2812	1982	10
2814	2044	10
2816	2030	10
2818	2214	10
2820	2040	10
2822	2192	10
2824	1992	10
2826	1976	10
2828	2048	10
2830	2216	10
2834	2218	10
2836	2222	10
2838	1988	10
2840	2234	10
2880	2182	14
2904	2178	14
2918	2184	14
2924	2098	14
2938	2100	14
2942	2244	10
2944	2036	10
2946	1990	10
2948	2240	10
2950	2220	10
2956	2236	10
2958	2248	10
2960	2046	10
2962	2208	10
2964	2252	10
2966	2246	10
2968	2238	10
2970	2206	10
2974	2242	10
2976	2250	10
3002	2210	10
3004	1928	10
3006	2202	10
3018	2194	10
3050	2270	14
3052	2266	14
3056	2264	14
3058	2262	14
3060	2260	14
3068	2254	14
3074	2258	14
3082	2224	10
3084	2226	10
3086	2232	10
3090	2256	14
3108	2228	14
3118	2230	14
3166	2200	10
4114	3944	14/16
4230	3906	10
4238	3910	10
4357	3876	20
4361	3878	20
4363	3880	20
4369	3936	14/18
4371	3890	14
4373	3898	16
4375	3848	14
4379	3888	16
4381	3886	14
4383	3850	14
4385	3894	14
4387	3904	10
4389	3908	10
4391	3976	10
4393	3914	10
4395	3978	10
4397	3980	10
4399	3916	10
4401	3912	10
4403	4016	10
4405	3920	10
4407	3946	14
4409	3896	14
4411	3852	14
4413	3884	14
4415	3938	14/18
4419	3950	14/18
4423	3942	14
4425	3940	14
4427	4004	14
4429	4014	14
4431	4036	14
4433	3954	14
4435	3902	14
4437	3948	14
4439	3898	14
4441	3998	14
4443	3952	16
4447	4000	14
4448	3900	14
4449	4042	14
4451	4010	14
4453	4006	14
4455	4002	14
4457	4012	14
4459	4048	16
4461	4068	14
4463	3996	14
4465	4050	14
4467	4054	14
4469	4070	14
4471	4072	14/18
4473	4076	14
4475	4074	14
4477	4044	14
4479	4106	14/18
4481	4008	14
4483	4066	14/18
4485	4078	14/18
4487	4156	20/25
4489	4080	20
4491	4056	20
4497	3982	10
4499	3922	10
4501	3918	10
4503	4086	10
4505	3984	10
4507	4020	10
4509	4018	10
4511	4022	10
4513	3986	10
4515	3990	10
4517	4104	14
4519	4046	16
4521	4052	14
4523	4080	14
4525	4084	16
4527	4112	14
4529	4082	14
4531	4114	14/18
4533	4040	14
4535	4038	14

4537	4144	30/35
4541	4288	30/35
4543	4150	30/35
4545	4148	30/35
4549	4058	20
4551	4062	20
4553	4064	20
4555	4060	20
4559	4178	14/18
4561	4182	14/18
4563	4180	14/18
4575	4176	14/18
4579	3994	10
4581	3988	10
4582	4030	10/12
4583	4026	10/12
4585	4024	10
4587	4028	10/12
4591	4418	10/12
4593	4032	10/12
4595	4424	10
4597	4422	10
4599	4124	10/12
4601	4130	12
4603	4126	12/14
4605	4128	12/14
4607	4136	12/14
4609	4132	12/14
4611	4134	12/14
4613	4138	12/14
4615	4142	12/14
4617	4140	12/14
4619	4108	14/18
4621	4110	16/20
4623	4234	16/20
4625	4116	16/20
4627	4228	16/20
4629	4184	14/18
4631	4206	14/18
4633	4214	14/18
4635	4218	14/18
4641	4220	14/18
4643	4208	14/18
4645	4212	14/18
4647	4210	14/18
4649	4090	20/25
4651	4162	20/25
4653	4160	20/25
4655	4164	20/25
4657	4292	20/25
4659	4240	14/18
4661	4246	16
4665	4216	14/18
4667	4168	14
4669	4166	14/18
4671	4172	14/18
4673	4174	14/18
4675	4316	14/18
4677	4170	14/18
4679	4250	12/14
4681	4248	12/14
4683	4258	12/14
4685	4252	12/14
4687	4254	12/14
4689	4256	12/14
4691	4262	12/14
4693	4260	12/14
4695	4264	12/14
4697	4266	12/14
4699	3882	20/25
4709	4280	12/14
4711	4274	12/14
4713	4268	12/14
4715	4272	12/14
4717	4276	12/14
4719	4286	12/14
4722	4278	12/14
4723	4270	12/14
4725	4284	12/14
4727	4282	12/14
4729	4236	16/20
4731	4362	16/20
4733	4356	16/20
4735	4120	16/20
4737	4224	16/20
4739	4118	16/20
4741	4230	16/20
4743	4242	16/20
4745	4368	16/20
4747	4366	16/20
4749	4350	14/18
4751	4354	14/18
4753	4318	14/18
4755	4308	14/18
4757	4302	14/18
4759	4226	14/18
4761	4306	14/18
4763	4482	14/18
4765	4706	14/18
4767	4708	14/18
4769	4392	12/14
4771	4336	12/14
4773	4390	12/14
4775	4324	12/14
4777	4326	12/14
4779	4322	12/14
4781	4380	12
4783	4334	12/14
4785	4388	12/14
4787	4320	12/14
4789	4384	12/14
4791	4396	12/14
4793	4394	12/14
4795	4382	12
4797	4338	12/14
4799	4328	12/14
4801	4398	12/14
4803	4386	12/14
4805	4330	12
4807	4434	12/14
4809	4430	12/14
4811	4442	12/14
4813	4440	12/14
4815	4446	12/14
4817	4448	12/14
4819	4460	12/14
4821	4464	12/14
4823	4332	12
4825	4438	12/14
4827	4444	12/14
4829	4232	16/20
4831	4244	16/20
4833	4238	16/20
4835	4312	16/20
4837	4310	16/20
4839	4372	16/20
4841	4304	16/20
4843	4360	16/20
4845	4364	16/20
4847	4484	16/20
4849	4298	25/30
4851	4450	25/30
4853	4452	25/30
4855	4454	25/30
4857	4296	25/30
4859	4462	12/14
4861	4432	12/14
4863	4468	12/14
4865	4472	12/14
4867	4474	12/14
4869	4466	12/14
4871	4476	12/14
4873	4436	12/14
4875	4500	12/14
4877	4504	12/14
4879	4502	12/14
4881	4506	12/14
4883	4508	12/14
4885	4478	12/14
4887	4514	12/14
4889	4512	12/14
4891	4536	12/14
4893	4518	12/14
4895	4516	12/14
4899	4520	25/30
4901	4458	25/30
4903	4456	25/30
4905	4522	25/30
4907	4092	24/30
4909	4530	12/14
4911	4534	12/14
4913	4532	12/14
4915	4540	12/14
4917	4544	12/14
4919	4538	12/14
4921	4542	12/14
4923	4510	12/14
4925	4564	12/14
4927	4546	12/14
4929	4410	10/12
4931	4402	10/12
4933	4416	10/12
4935	4036	14
4937	4414	10/12
4939	4404	10/12
4941	4412	10/12
4945	4846	10/12
4947	4710	10/12
4949	4568	12/14
4951	4576	12/14
4953	4566	12/14
4955	4570	12/14
4957	4572	12/14
4959	4562	12/14
4961	4612	12/14
4963	4548	12/14
4965	4602	12/14
4967	4560	12/14
4969	4486	16/20
4971	4370	16/20
4973	4376	16/20
4975	4480	16/20
4977	4494	16/20
4979	4498	16/20
4981	4488	16/20
4983	4374	16/20
4985	4496	16/20
4987	4378	16/20
4989	4550	25/30
4991	4526	25/30
4993	4528	25/30
4995	4620	25/30
4997	4524	25/30
4999	4552	25/30
5001	4622	25/30
5003	4096	25/30
5005	4554	25/30
5007	4624	25/30
5009	4574	12/14
5011	4614	12/14
5013	4608	12/14
5015	4658	12/14
5019	4610	12/14
5021	4618	12/14
5023	4600	12/14
5025	4604	12/14
5029	4640	12/14
5031	4652	12/14
5033	4616	12/14
5035	4642	12/14
5037	4644	12/14
5039	4646	12/14
5041	4648	12/14
5043	4650	12/14
5045	4654	12/14
5047	4686	12/14
5049	4660	16/20
5051	4664	16/20
5053	4666	16/20
5055	4670	16/20
5057	4684	16/20
5059	4662	16/20
5061	4490	16/20
5063	4674	16/20
5065	4734	16/20
5069	4656	12/14
5071	4696	12/14
5073	4704	12/14
5075	4746	12/14
5077	4694	12/14
5079	4692	12/14
5081	4702	12/14
5083	4688	12/14
5085	4690	12/14
5087	4698	12/14
5091	4732	14/18
5093	4680	14/18
5095	4768	14/18
5101	4718	14/18
5111	4726	16/20
5113	4672	16/20
5115	4676	16/20
5117	4730	16/20
5119	4736	16/20
5121	4668	16/20
5123	4740	16/20
5125	4774	16/20
5127	4578	12/14
5129	4738	16/20
5131	4760	12/14
5133	4748	12/14
5137	4700	12/14
5141	4756	12/14
5149	4786	12/14
5151	4742	16/20
5153	4712	16/20
5155	4728	16/20
5157	4780	16/20
5159	4714	16/20
5161	4848	12
5163	4788	12/14
5169	4828	12/14
5177	4800	12
5181	4762	12/14
5183	4804	12
5185	4758	12/14
5187	4764	12/14
5189	4832	12/14
5191	4840	12/14
5193	4834	12/14
5195	4798	12
5197	4838	12/14
5199	4830	12/14
5201	4806	16/20
5203	4806	16/20
5205	4778	16/20
5207	4744	16/20
5209	4818	16/20
5211	4824	16/20
5213	4784	16/20
5215	4868	16/20
5217	4770	16/20
5219	4822	16/20
5221	4556	25/30
5223	4628	25/30
5227	4558	25/30
5229	4630	25/30
5231	4716	16/20
5233	4816	16/20
5235	4766	16/20
5237	4812	16/20
5239	4814	16/20
5241	4420	10
5243	4406	10/12
5245	4408	10/12
5247	4186	10/12
5261	4852	12/14
5263	4836	12/14
5265	4858	12/14
5267	4854	12/14
5269	4850	12
5271	4862	12/14
5273	4826	12/14
5275	4920	12/14
5277	4842	12/14
5279	4922	12/14
5281	4884	16/20
5283	4872	16/20
5285	4782	16/20
5289	4810	16/20
5291	4820	16/20
5295	4886	16/20
5297	4892	16/20
5299	4874	16/20
5301	4878	16/20
5303	4722	16/20
5305	4896	16/20
5309	4724	16/20
5311	4870	16/20
5313	4898	16/20
5317	4900	16/20
5321	4906	16/20
5323	4880	16/20
5325	4904	16/20
5327	4882	16/20
5339	4750	12/14
5341	4924	12/14
5343	4856	12/14
5345	4926	12/14
5349	4860	12/14
5351	4928	12/14
5353	4918	12
5355	4844	12/14
5357	4930	12/14
5361	4932	12/14
5365	4910	16/20
5367	4908	16/20
5369	4902	16/20
5371	4912	16/20
5373	4720	16/20
5375	4894	16/20
5381	4876	16/20
5397	4794	12/14
5401	4470	12/14
5405	4802	12/14
5407	4792	12/14
5415	4632	25/30
5417	4636	25/30
5419	4634	25/30
5434	4606	12/14
5445	4790	12
5447	4864	12/14
5459	4796	12/14
5465	4916	12/14
5467	4866	12/14
5543	4752	12/14
5545	4754	12/14
5599	4428	10

ALLDAYS & ONIONS
PNEUMATIC ENGINEERING CO. LTD.

PATENTS

This section contains illustrations taken from some of the patents taken out by the company which may assist the reader to understand how particular products were developed and also helps to illustrate, again, the enormous diversity of products manufactured by Alldays & Onions.

The drawings, of course, are not to the same scale and are presented in product groups rather than in true chronological order.

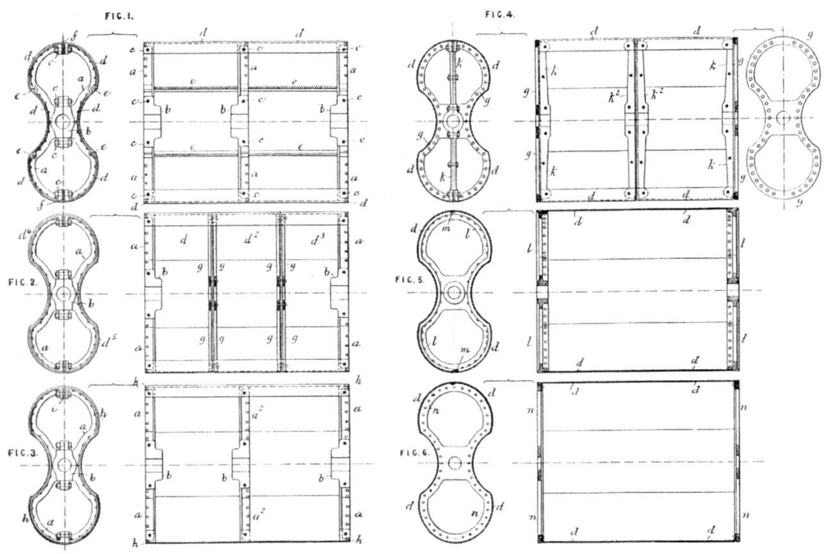
FIG. 1.
FIG. 2.
FIG. 3.
FIG. 4.
FIG. 5.
FIG. 6.

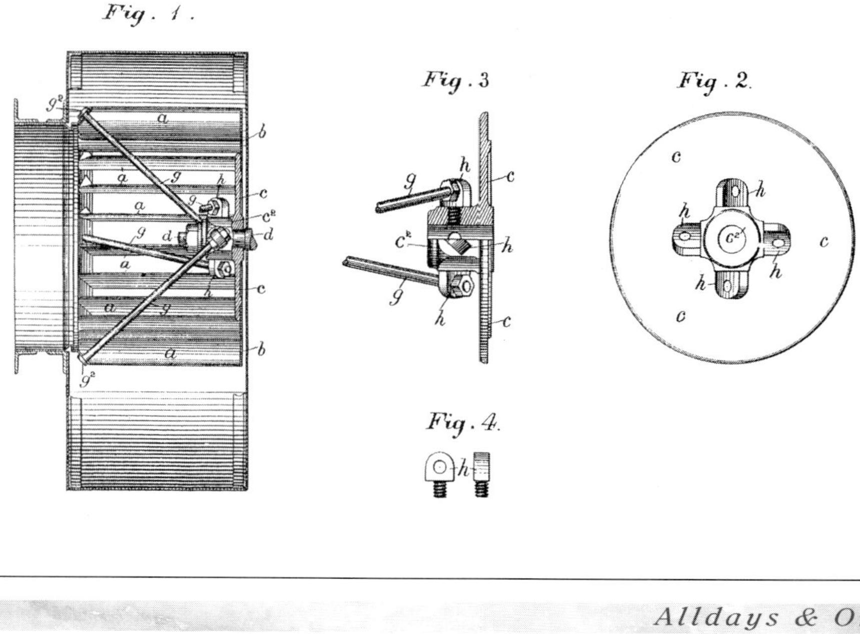
Fig. 1.
Fig. 3
Fig. 2.
Fig. 4.

LEFT-TOP: PATENT NUMBER 5919.
Taken out jointly by William Allday (the younger) and Edmund Allday in December 1883 for improvements in the design of revolvers, or rotors, as used in Root's type rotary blowers and exhausters.

LEFT-BOTTOM: PATENT NUMBER 10845.
Taken out in October 1906 by Edward Livingstone Joselin, described as a Mechanical Engineer at the Great Western Works, relates to improvements in the construction of large fans of the centrifugal type

BELOW: PATENT NUMBER 16476.
Also taken out by Edward Livingstone Joselin, in December 1909, this patent covers the design of compact centrifugal fans enclosed in cast iron casings and driven directly by electric motors.

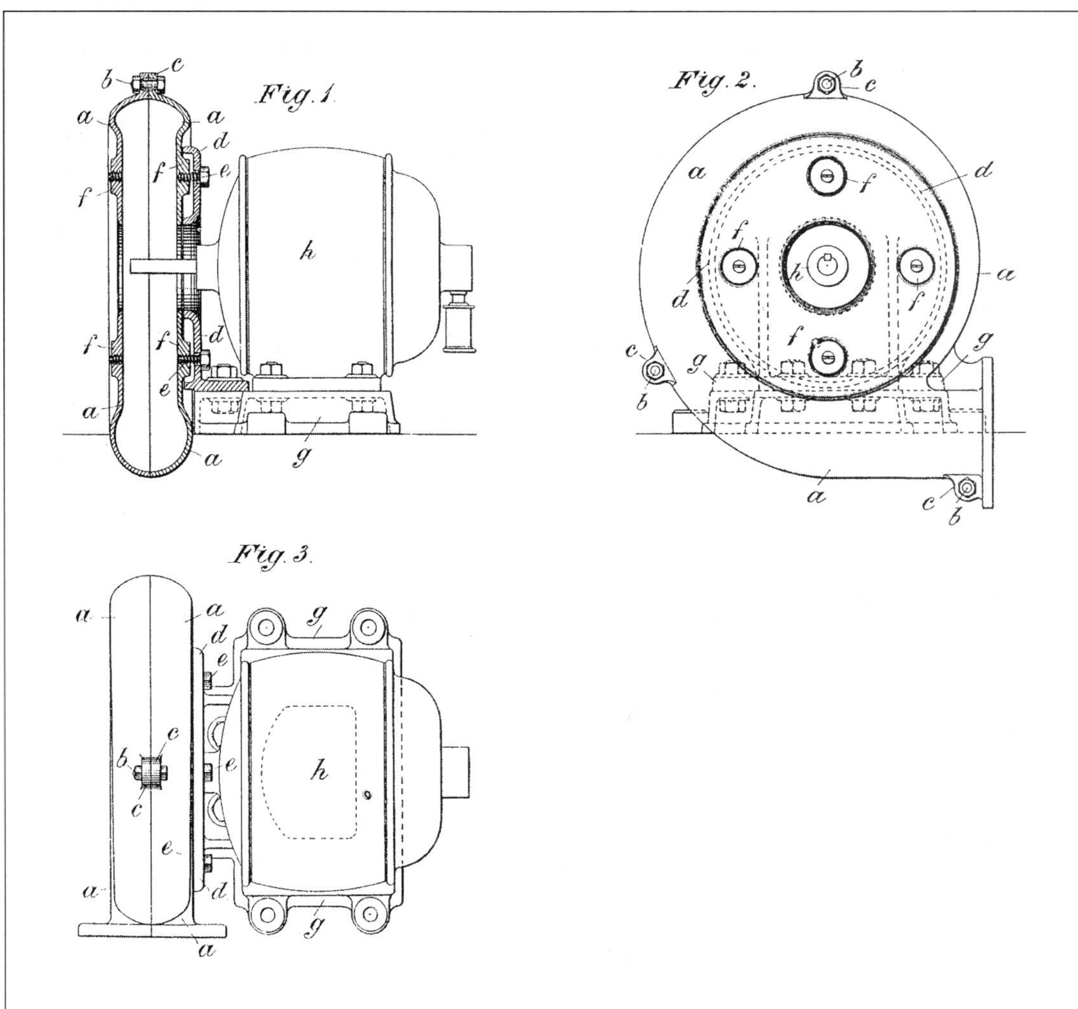

Fig. 2.

Fig. 1.

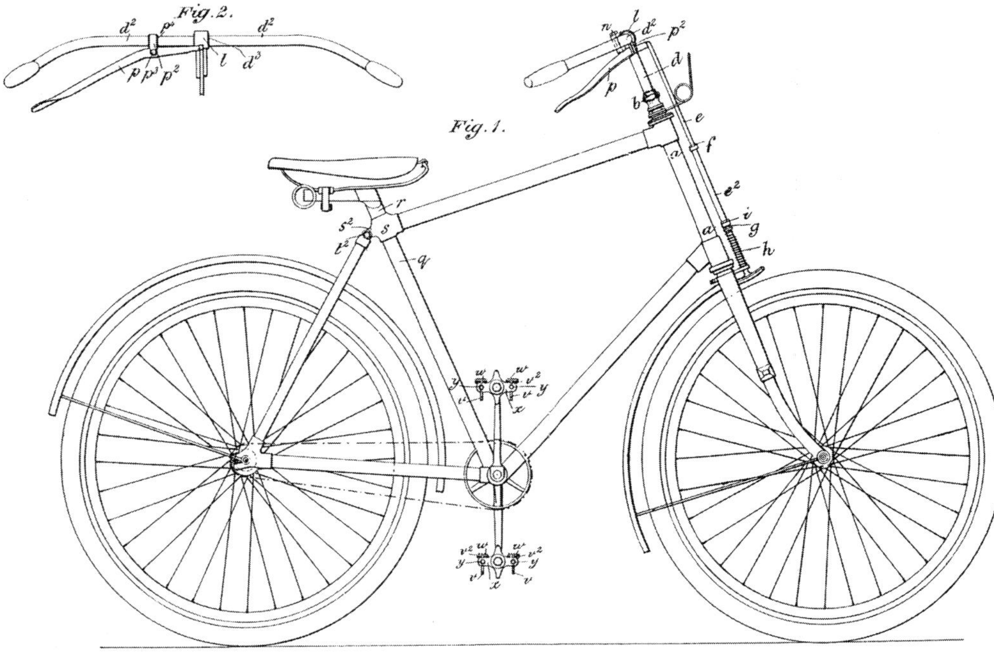

Fig. 24.

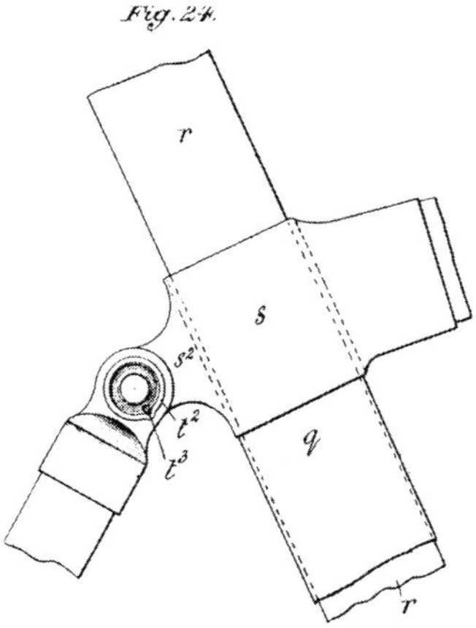

Fig. 26.

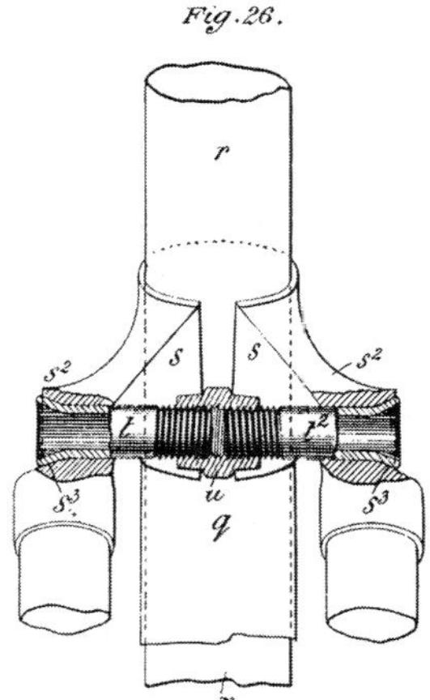

Fig. 29.

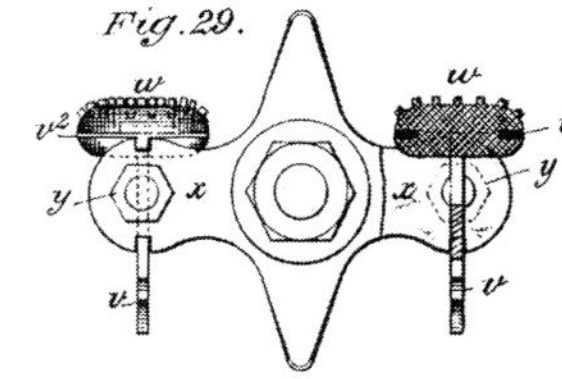

Fig. 27.

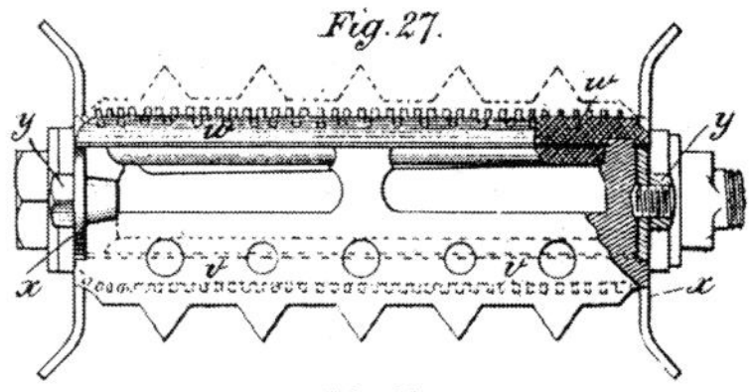

Fig. 28.

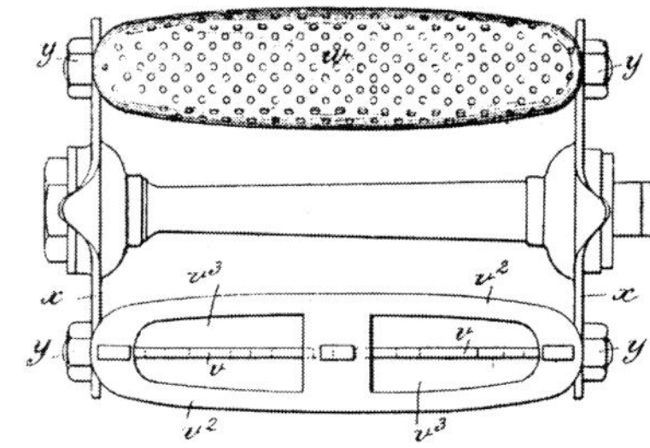

Fig. 25.

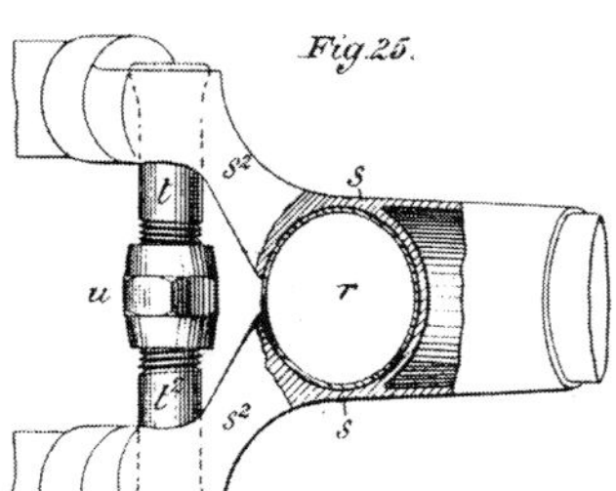

LEFT-TOP: PATENT NUMBER 16660.
Taken out jointly by William Allday and Edmund Allday in September 1892 covering improvements in the design of bicycles. This particular drawing relates to improvements in the design of the front wheel brake.

LEFT-BOTTOM: FROM THE SAME PATENT.
Details of the locking device for the adjustable seat tube and construction of pedals with rubber grips fitted to the footrests.

BELOW: PATENT NUMBER 18765.
Taken out jointly by William Allday and Edmund Allday in October 1893 covering several improvements in the design of bicycles, tricycles and other velocipedes, this particular drawing showing the construction of pedals and chain adjusters.

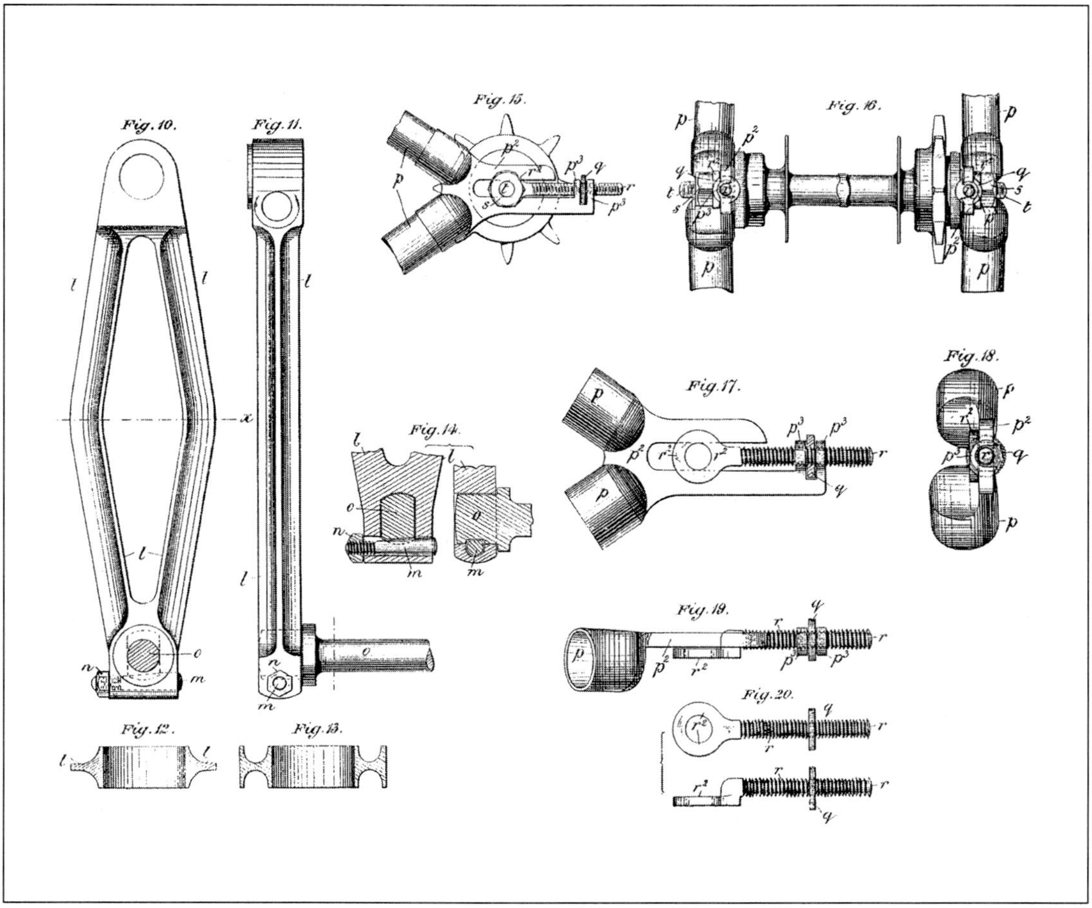

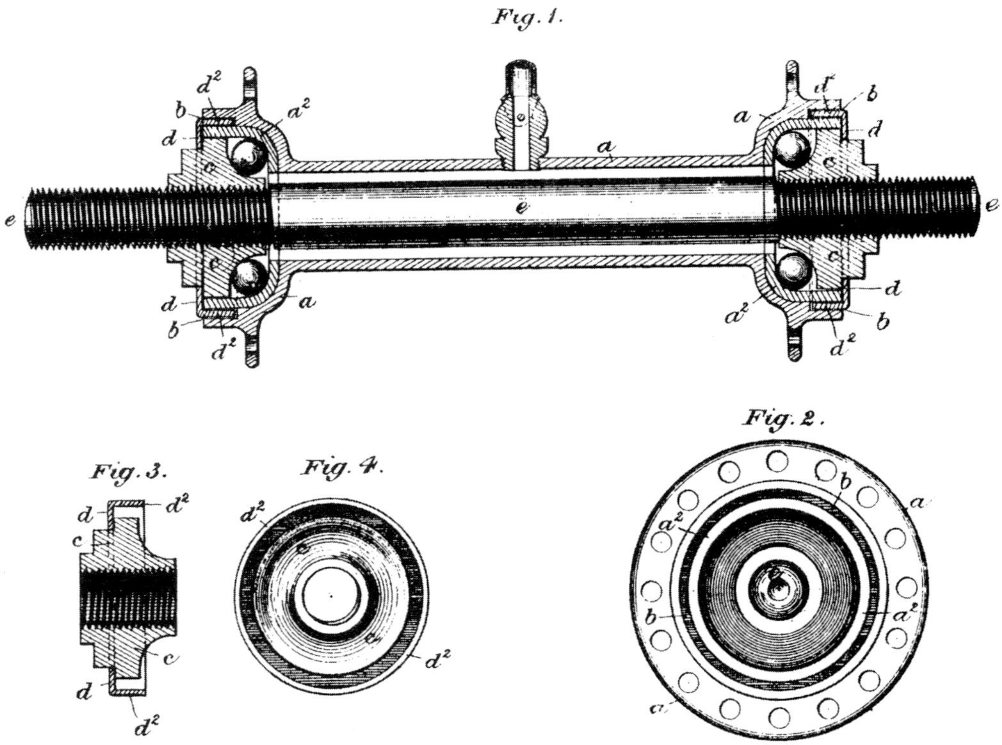
Fig. 1.
Fig. 2.
Fig. 3.
Fig. 4.

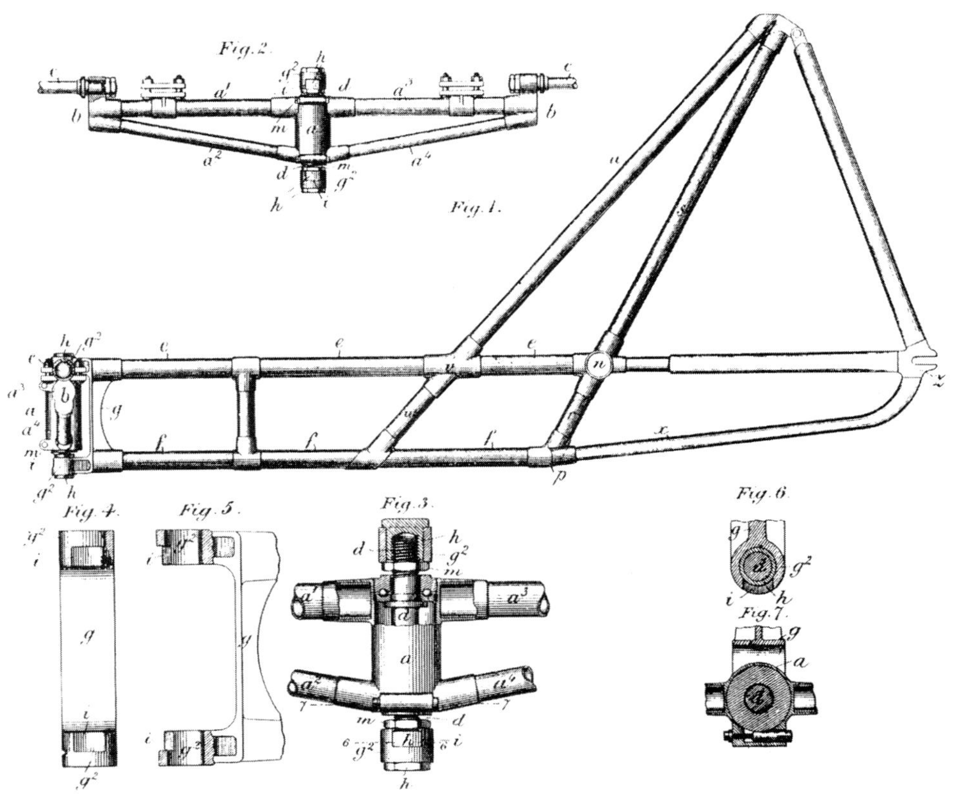
Fig. 2.
Fig. 1.
Fig. 4.
Fig. 5.
Fig. 3.
Fig. 6.
Fig. 7.

LEFT-TOP: PATENT NUMBER 16213.
Taken out jointly by William Allday and Edmund Allday in August 1895 for improvements in the construction to the front wheel bearing for bicycles, tricycles and other velocipedes.

LEFT-BOTTOM: PATENT NUMBER 3454.
Taken out in February 1909 by John Luckett, described as an Engineer at the Matchless Works, for improvements to the design of frames for tradesmen's carrier tricycles.

BELOW: PATENT NUMBER 365161.
Taken out jointly by Leigh Evans Clulee and Victor Earnest Barker in October 1930 this patent covers the design of a cycle trailer which allowed the cycle pulling it to be banked over on corners whilst keeping the two trailer wheels firmly in contact with the roadway.

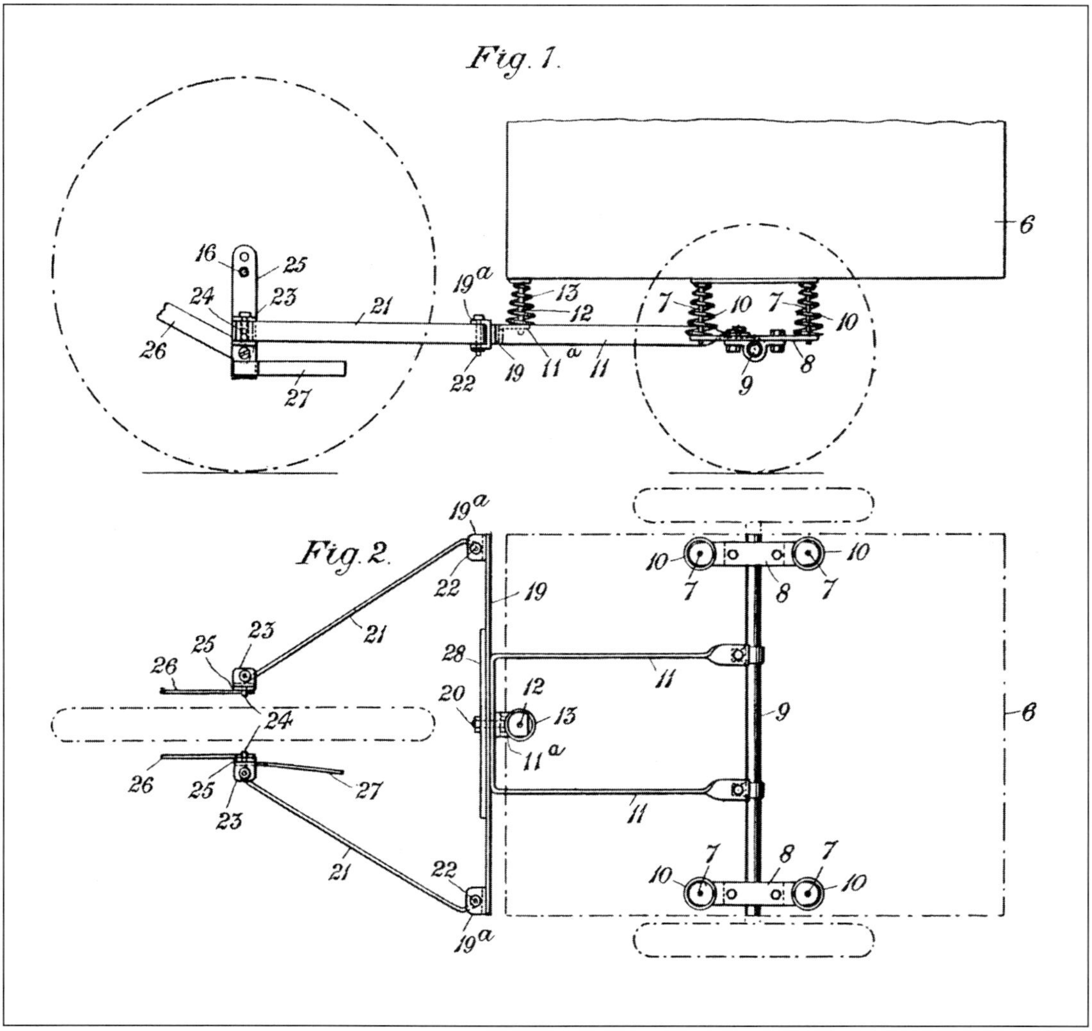

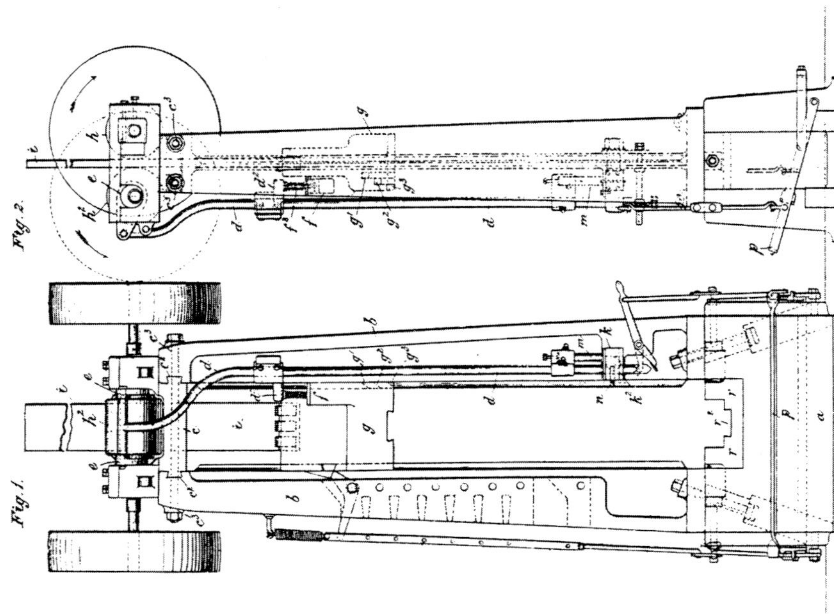

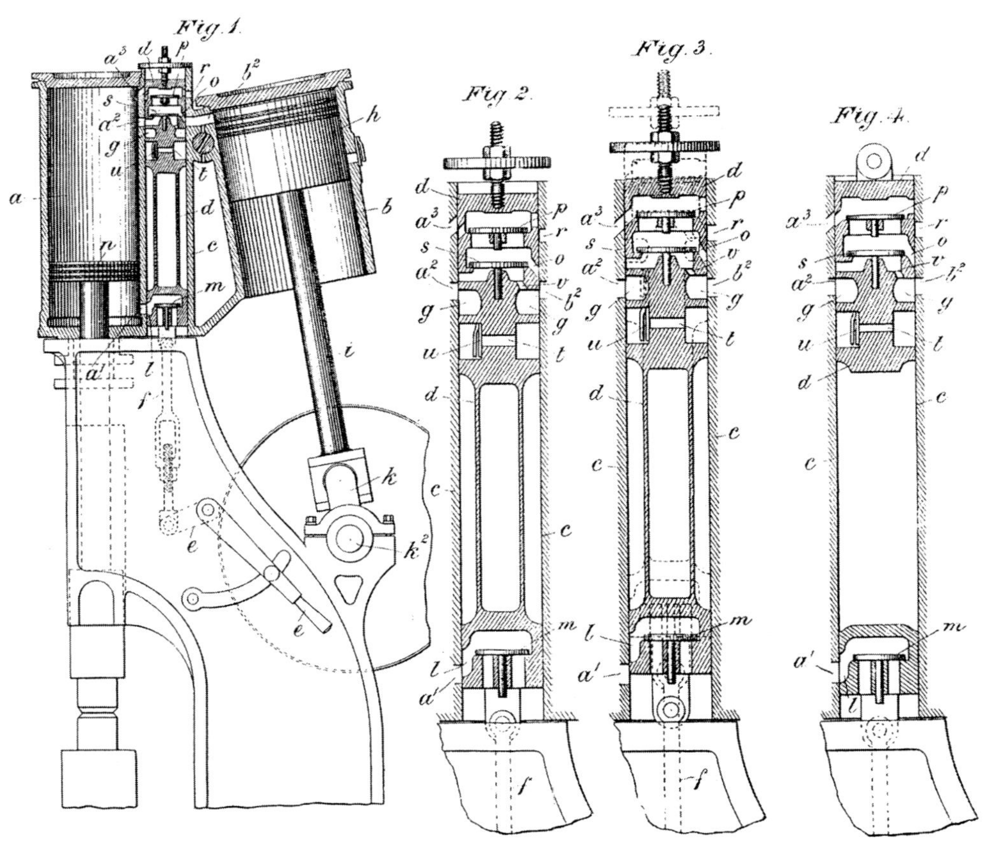
Fig. 1.
Fig. 2.
Fig. 3.
Fig. 4.

LEFT-TOP: PATENT NUMBER 11883.
Taken out in June 1900 by Robert Chandler Jennings, described as a Mechanical Engineer with the company, covers various improvements to drop hammers, or drop stamping machines used in the forging industry.

LEFT-BOTTOM: PATENT NUMBER 5508.
Taken out in March 1910 by Philip John Wilson, a Mechanical Engineer with the company. The patent covers improvements to forging hammers controlled by pneumatic cylinders.

BELOW: PATENT NUMBER 10831.
Another patent taken out by Philip John Wilson in May 1913, to cover improvements to drop stamp hammers.

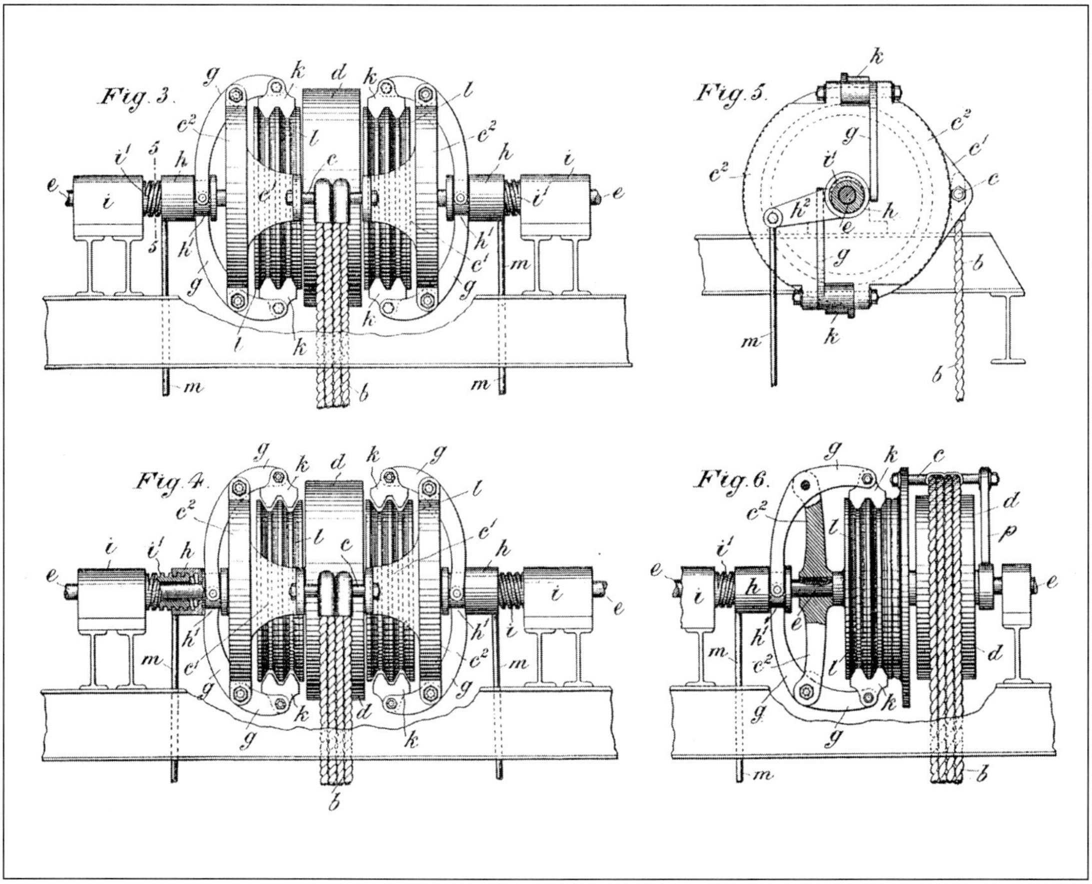

F I G. 4. F I G. 5.

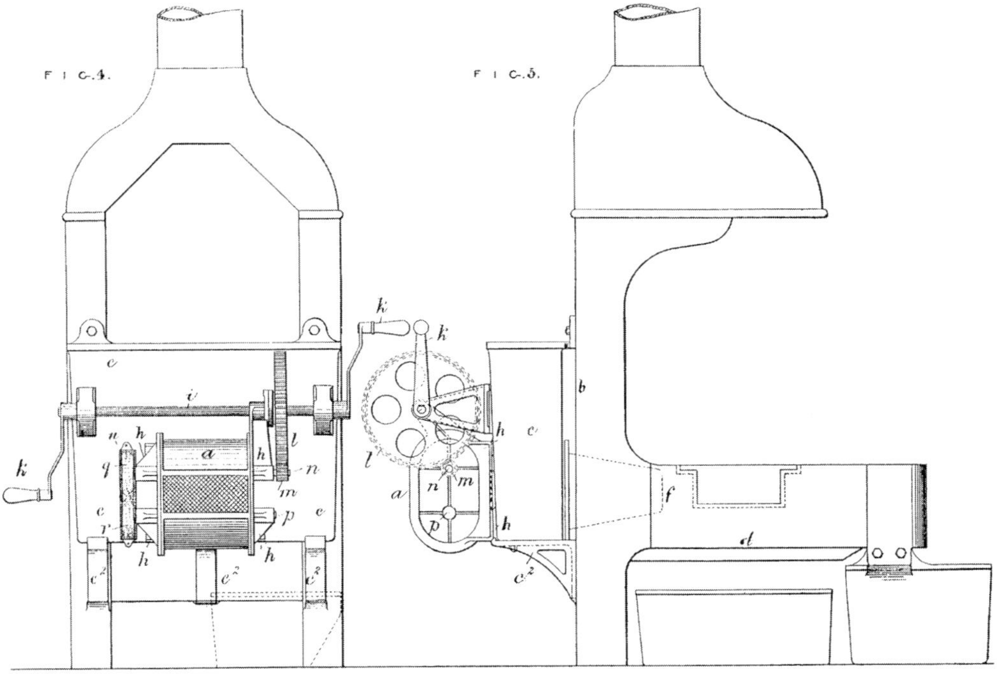

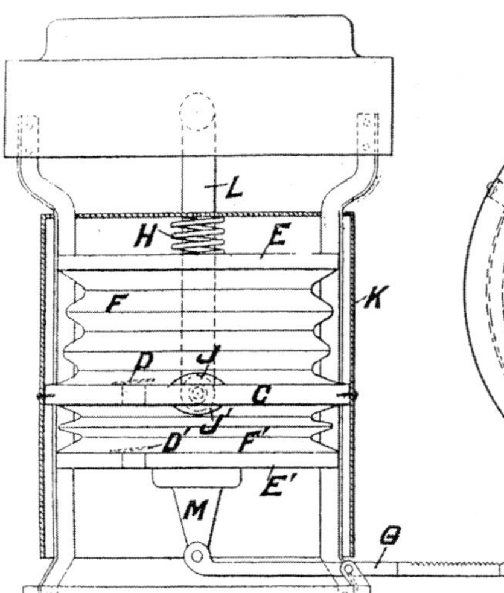

FIG. 1

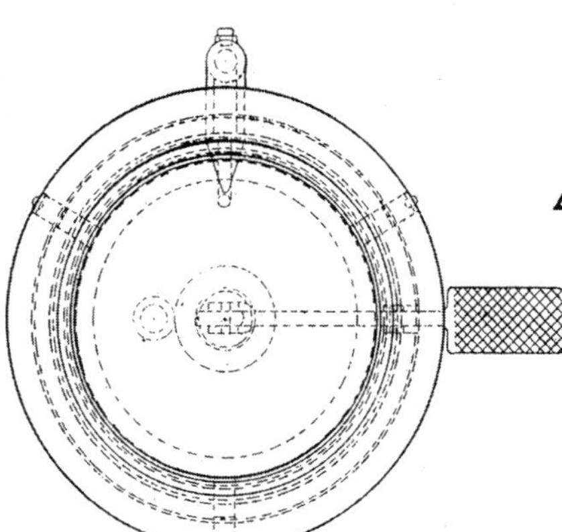

FIG. 2

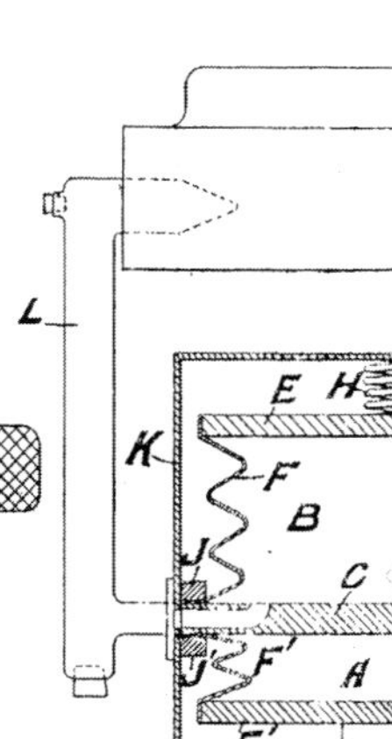

FIG. 3.

LEFT-TOP: PATENT NUMBER 2249.
Taken out in February 1886 by William Allday 'the younger', to cover improvements to portable forges fitted with hand operated Roots blowers.

LEFT-BOTTOM: PATENT NUMBER 396.
Taken out jointly by William Allday and Edmund Allday, in January 1894, for improvements to bellows used for portable and other forges.

BELOW: PATENT NUMBER 8632.
Taken out in April 1899 jointly by William Allday and Edmund Allday for improvements to hand operated bellows, with particular reference to prevent 'vermin' access to the interior of the bellows!

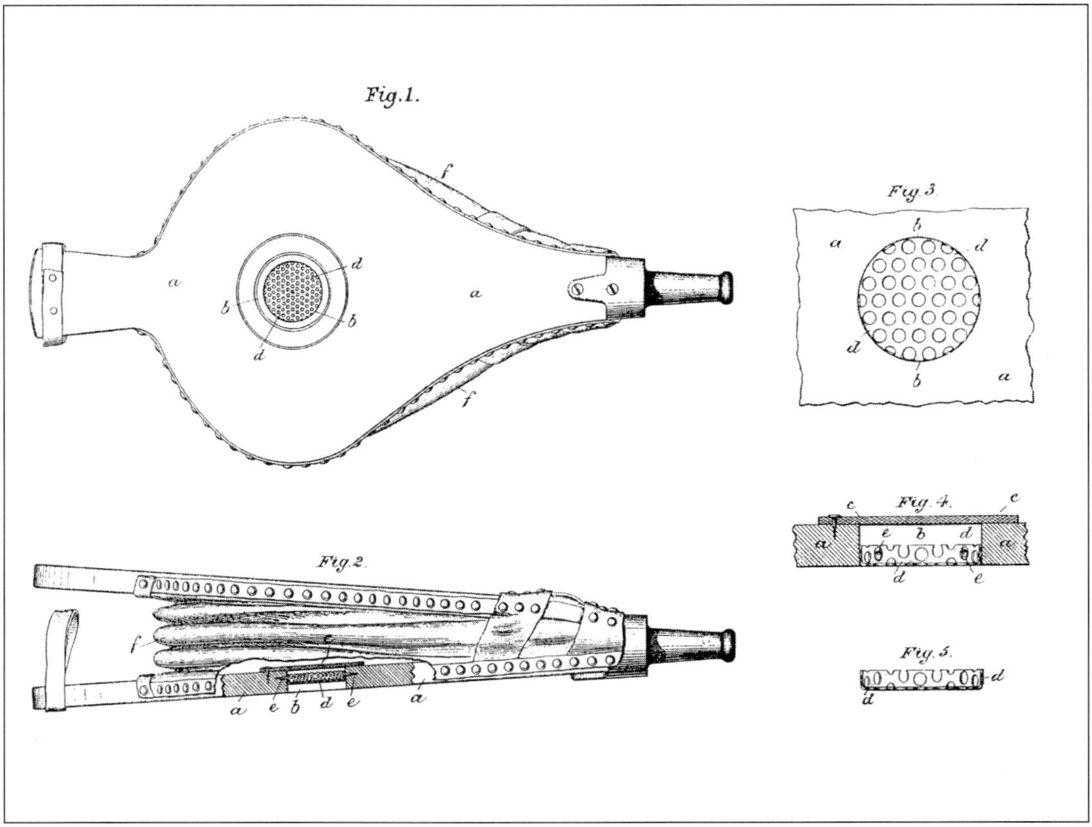

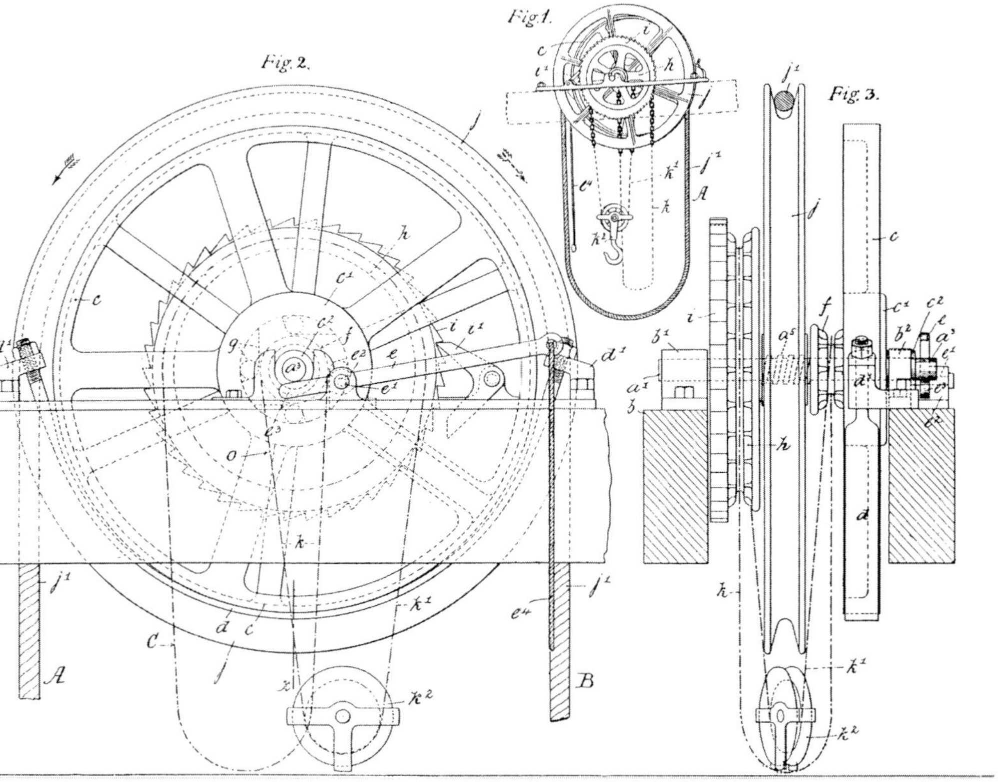
Fig. 1.
Fig. 2.
Fig. 3.
A
B

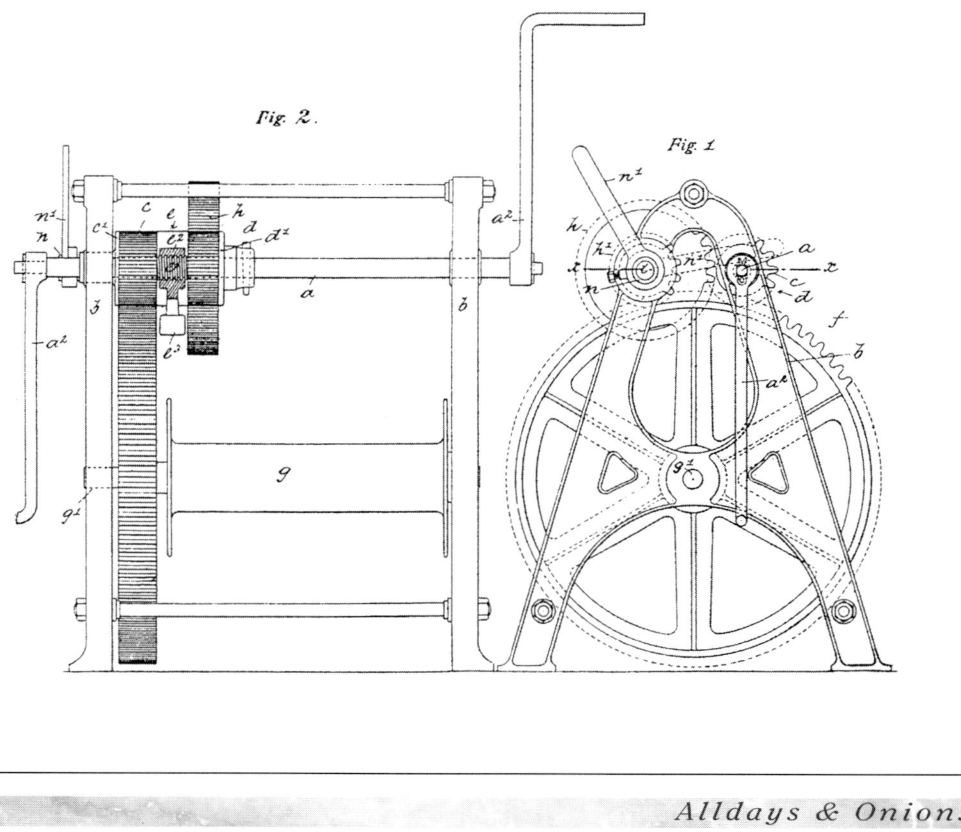
Fig. 2.
Fig. 1

LEFT-TOP: PATENT NUMBER 17377.
Taken out jointly by William Thomas Eades, Engineer, and Edmund Allday, also described as an Engineer. Dated August 1899 this patent covers improvements to hoists, particularly those used in warehouses.

LEFT-BOTTOM: PATENT NUMBER 23514.
Dated November 1899 and also taken out jointly by William Thomas Eades and Edmund Allday, this patent deals with the design of crab winches. Reference is also made in this patent to two other patents taken out in 1897 by Eades and Eades & Matthews for similar devices.

BELOW: PATENT NUMBER 4816.
Dated February 1914 this patent was taken out by Edmund Allday - Managing Director, John Newton Collins - Manager of the Fan Department, and Frank Lyon - Draughtsman, all of the Great Western Works. The patent covers the design of a novel truck for use in coal mines which distributed stone dust along tunnels to minimize the danger of explosions and fire.

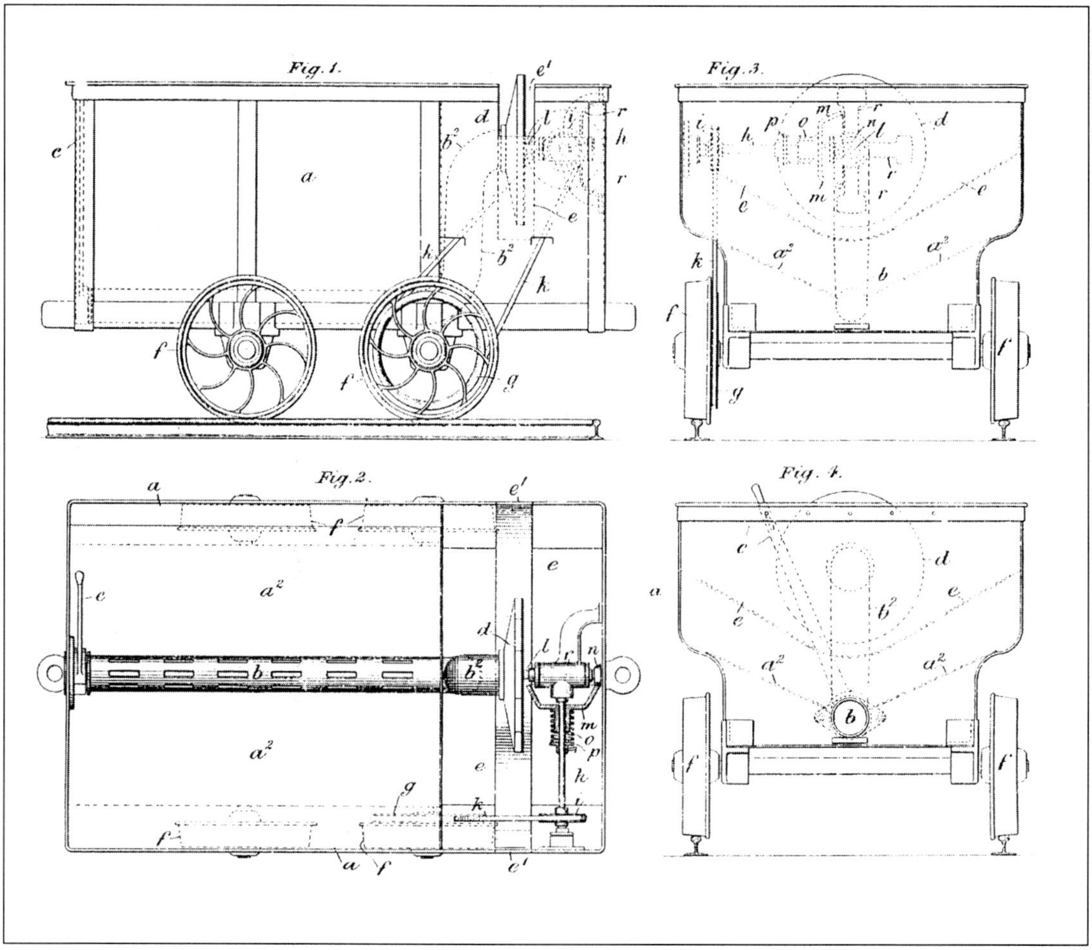

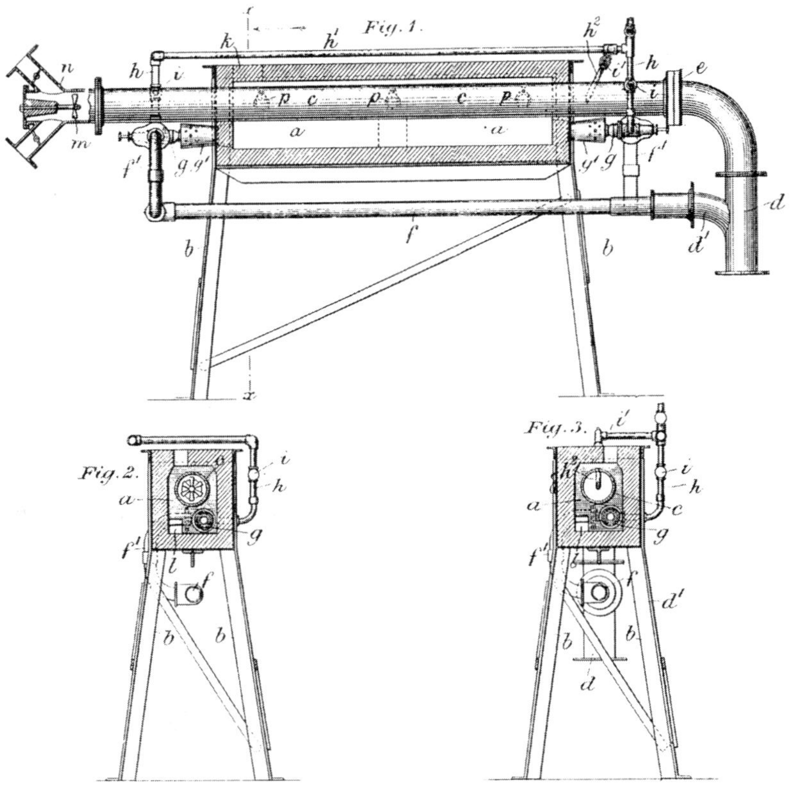
Fig. 1.
Fig. 2.
Fig. 3.

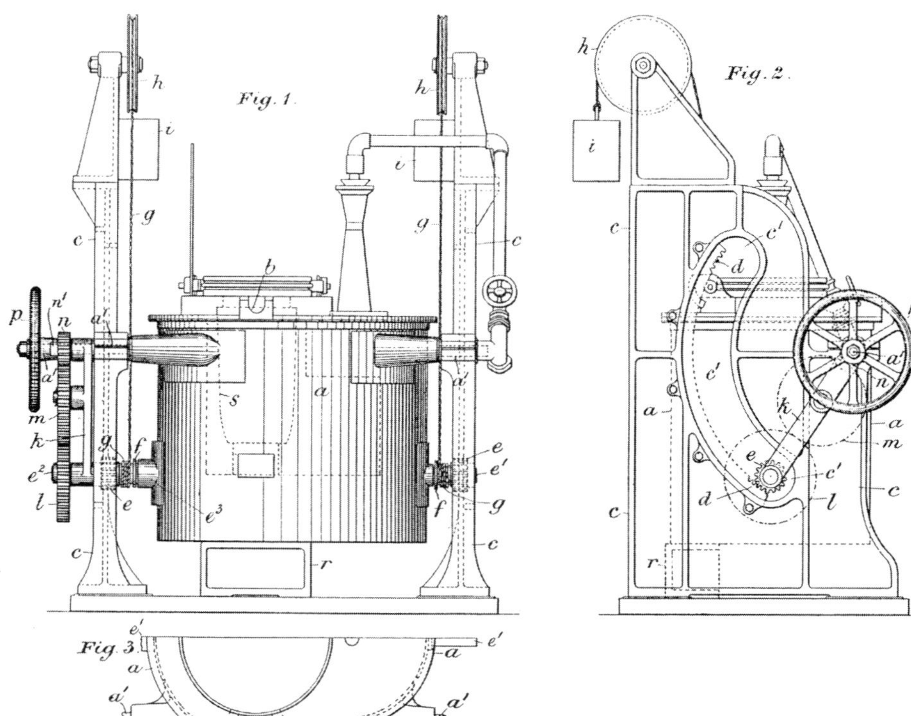
Fig. 1.
Fig. 2.
Fig. 3.

LEFT-TOP: PATENT NUMBER 113717.
Taken out in May 1917 by Percy William Allday describing improvements to devices for gasifying or vapourising liquid fuels, for furnaces, etc.

LEFT-BOTTOM: PATENT NUMBER 117212.
Dated November 1917 this patent was taken out jointly by Alexander Nicholson - Chief Draughtsman and Harold Hodkinson - Mechanical Draughtsman, both of Great Western Works, for improvements to the operating mechanism of tilting crucible furnaces.

BELOW: PATENT NUMBER 18335.
Taken out jointly by William Allday and Edmund Allday in September 1893 for improvements to engineers' bench vices.

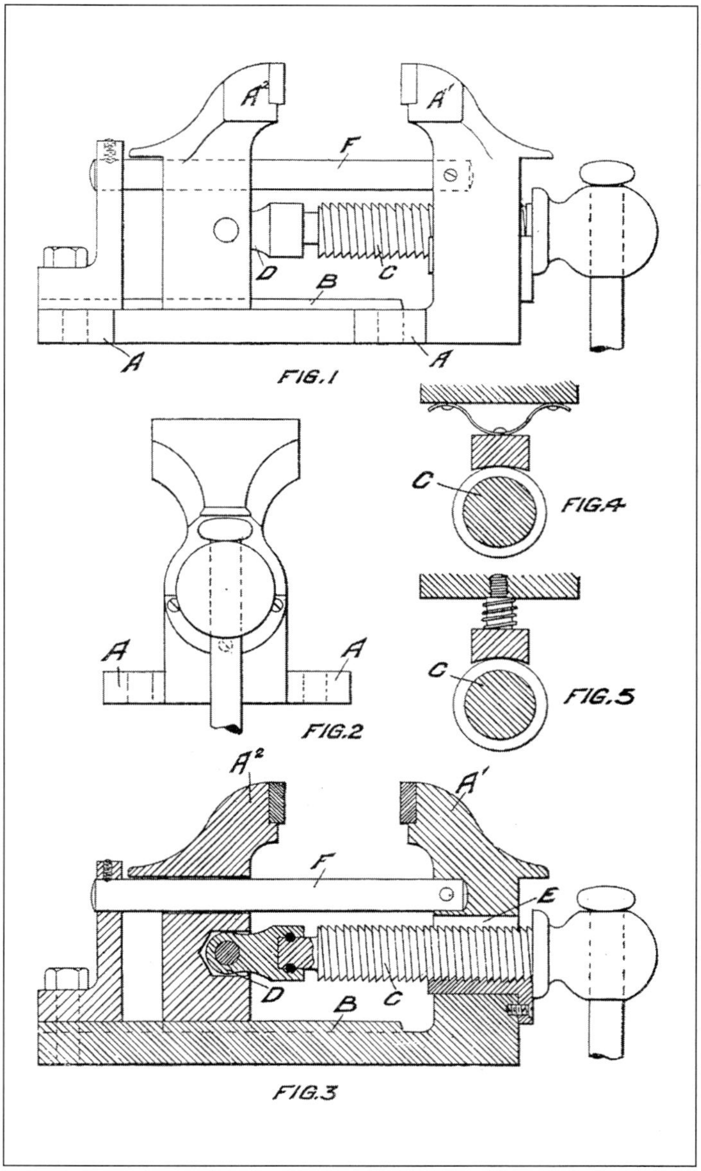

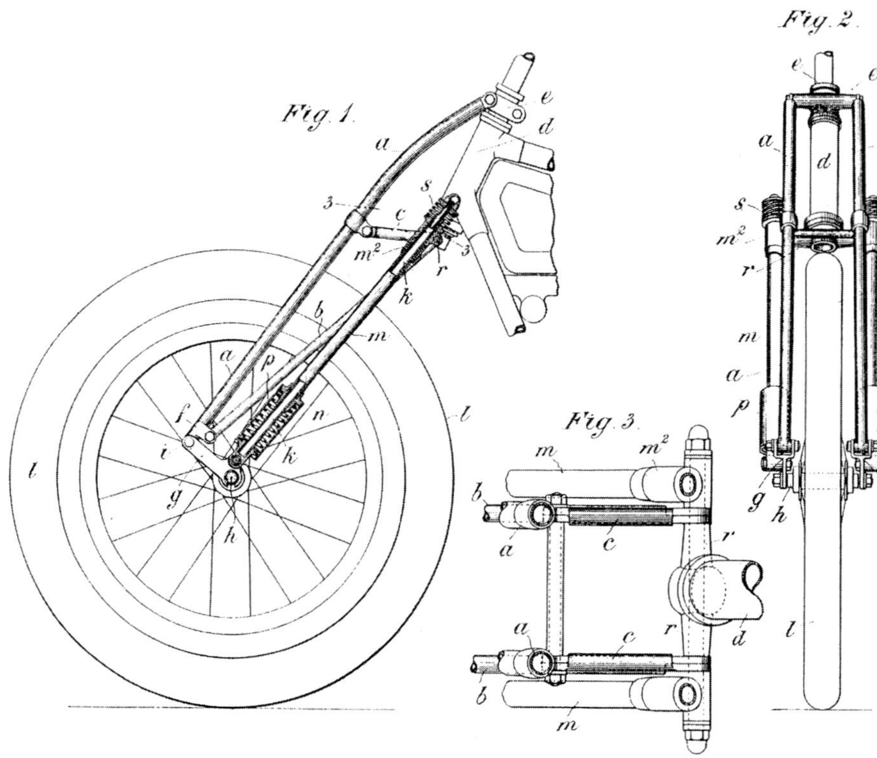
Fig. 1.
Fig. 2.
Fig. 3.

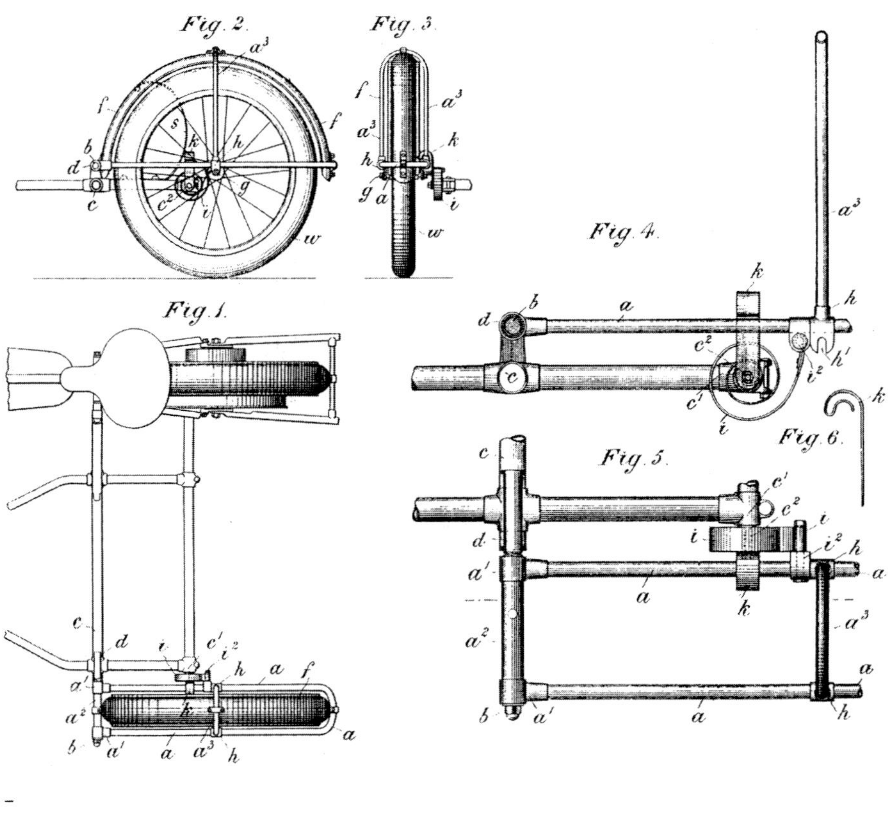
Fig. 2.
Fig. 3.
Fig. 4.
Fig. 1.
Fig. 5.
Fig. 6.

LEFT-TOP: PATENT NUMBER 15849.
Taken out in November 1915 by Albert Henry Johnson - Works Foreman of the Matchless Works, to cover the design of springing, or suspension, used on motor cycles.

LEFT-BOTTOM: PATENT NUMBER 120795.
Also taken out by Albert Henry Johnson, but in December 1917, this patent covers the design of springing for motor cycle side cars.

BELOW: PATENT NUMBER 122286.
A third patent by Albert Henry Johnson, taken out in February 1918, to cover the design of quick release couplings, or connectors, to attach side cars or trailers to cycles.

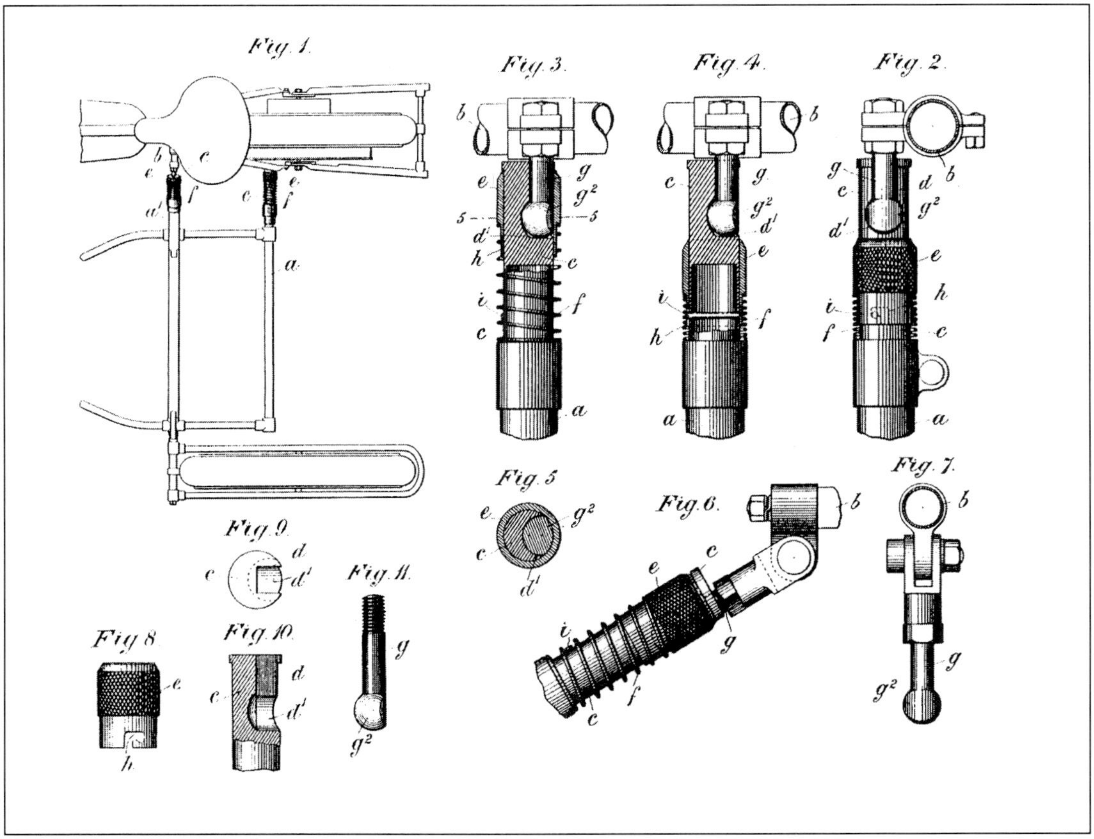

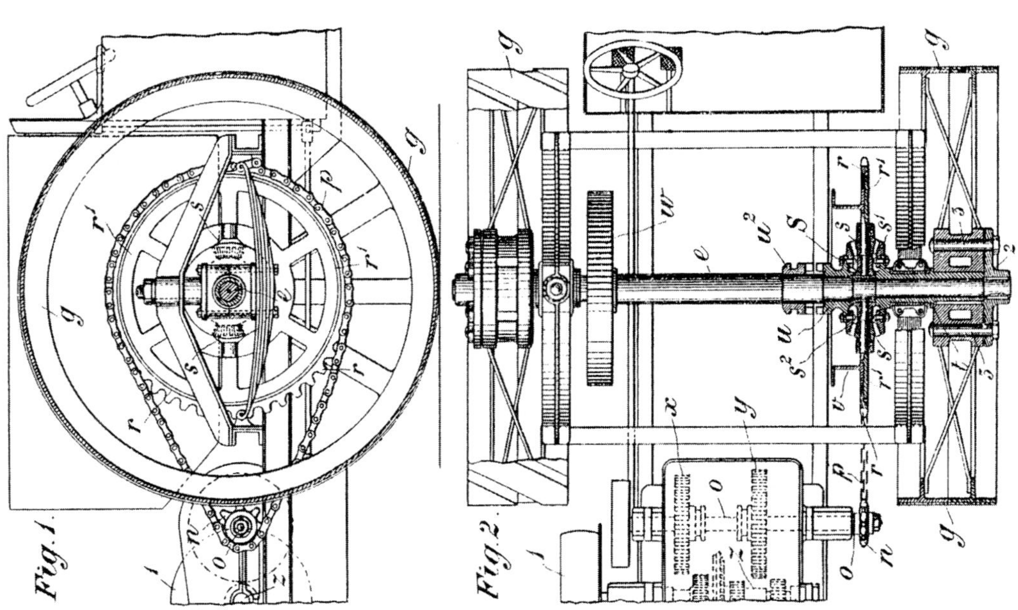
Fig.1.
Fig.2.

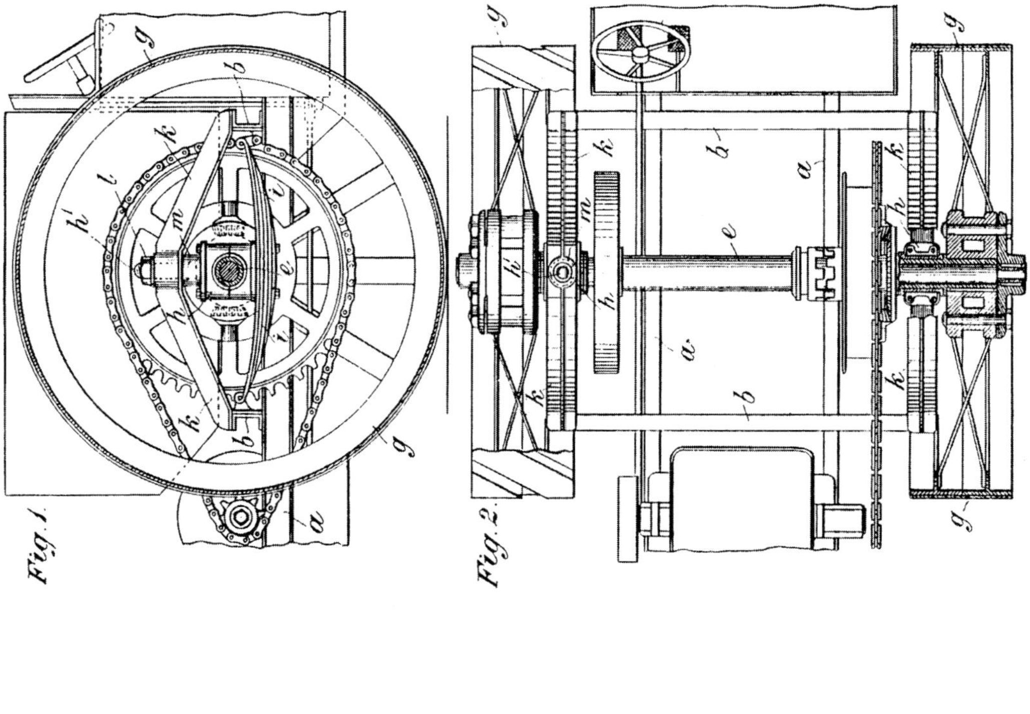
Fig.1.
Fig.2.

LEFT-TOP: PATENT NUMBER 115277.
Taken out jointly by Nathan Lees Hilton - General Manager, and Charles Edward Simms - Mechanical Engineer & Designer, both of the Matchless Works, in May 1917. This covers the design of improvements to the driving mechanism of motor tractors.

LEFT-BOTTOM: PATENT NUMBER 115398.
Also taken out jointly by Nathan Lees Hilton and Charles Edward Simms, in May 1917, to cover improvements in the construction and arrangement of parts used in motor tractors.

BELOW: PATENT NUMBER 116192.
A third patent taken out by Nathan Lees Hilton and Charles Edward Simms, in August 1917, dealing with the design of a chain operated steering mechanism suitable for a motor tractor.

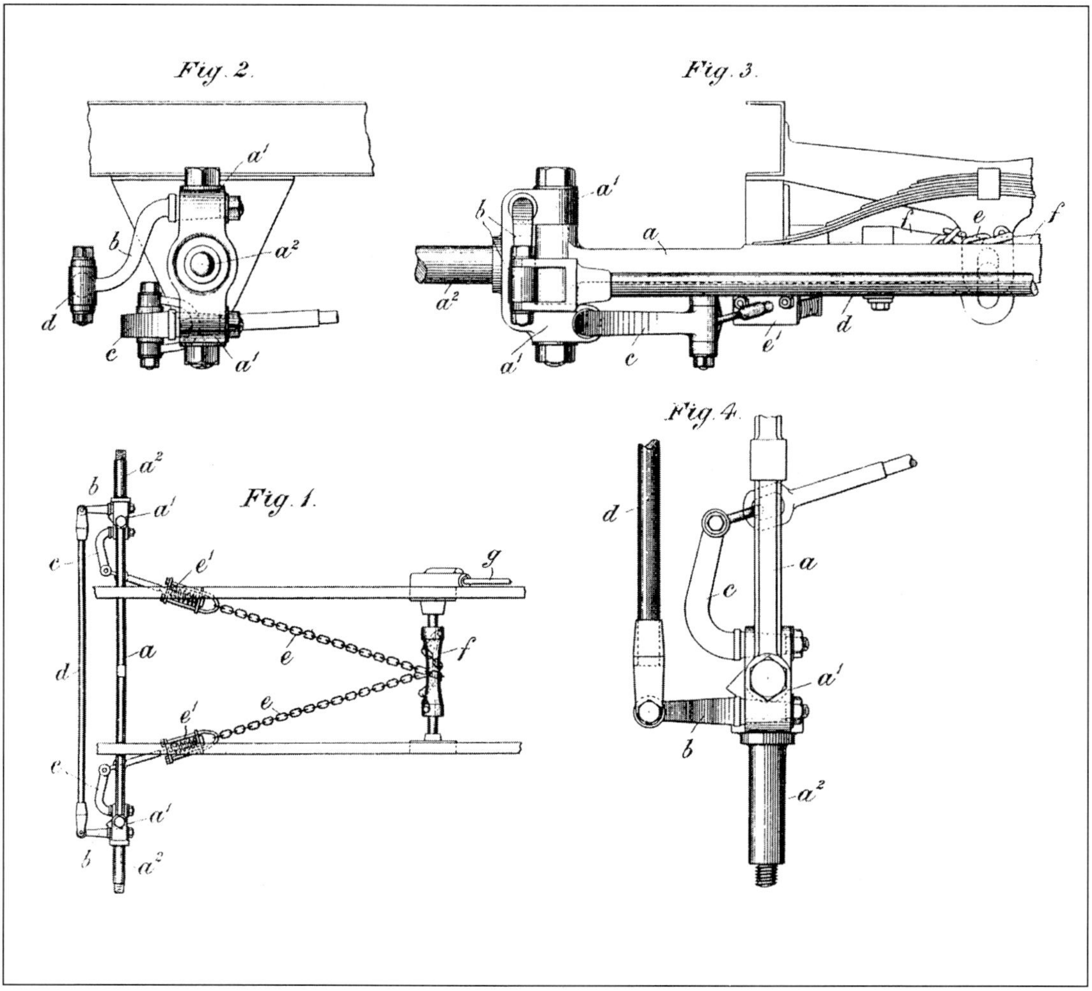

Fig. 1.

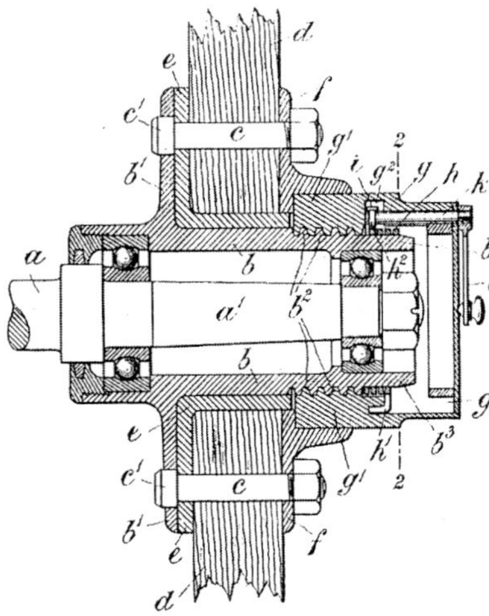

Fig. 2.

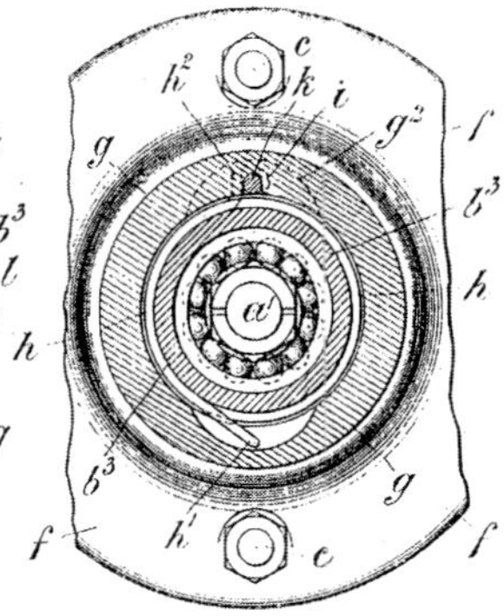

Fig. 3.

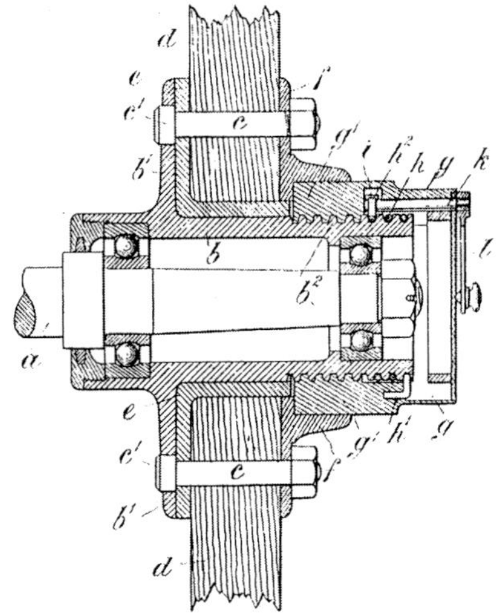

Fig. 1.

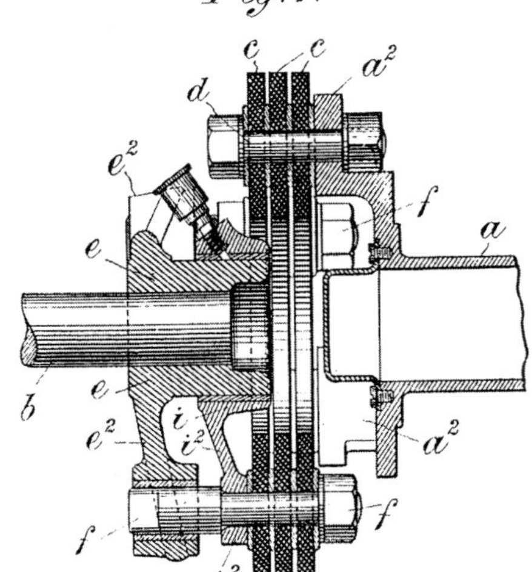

Fig. 2.

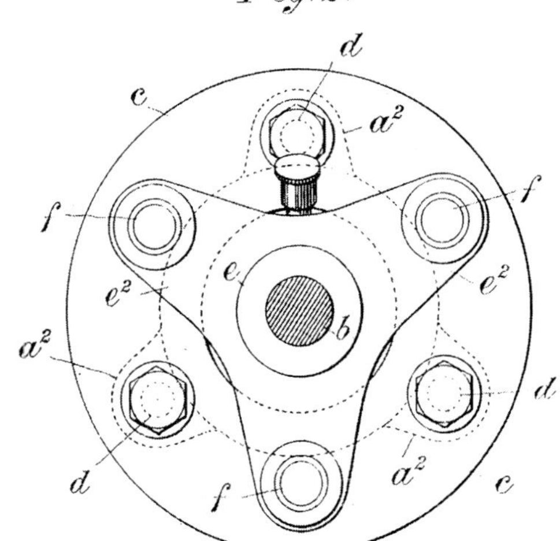

LEFT-TOP: PATENT NUMBER 22949.
Taken out in October 1909 by Charles Edward Simms, a Mechanical Engineer at the Matchless Works, for the design of detachable wheels suitable for motor cars and other vehicles.

LEFT-BOTTOM: PATENT NUMBER 125009.
Taken out in September 1918 by Arthur William Reeves - Manager of the Matchless Works, for an improved form of shaft coupling.

BELOW: PATENT NUMBER 150768.
Taken out in May 1919 by Charles Edward Simms for the design of a carburettor suitable for operating on paraffin, or alcohol on internal combustion engines.

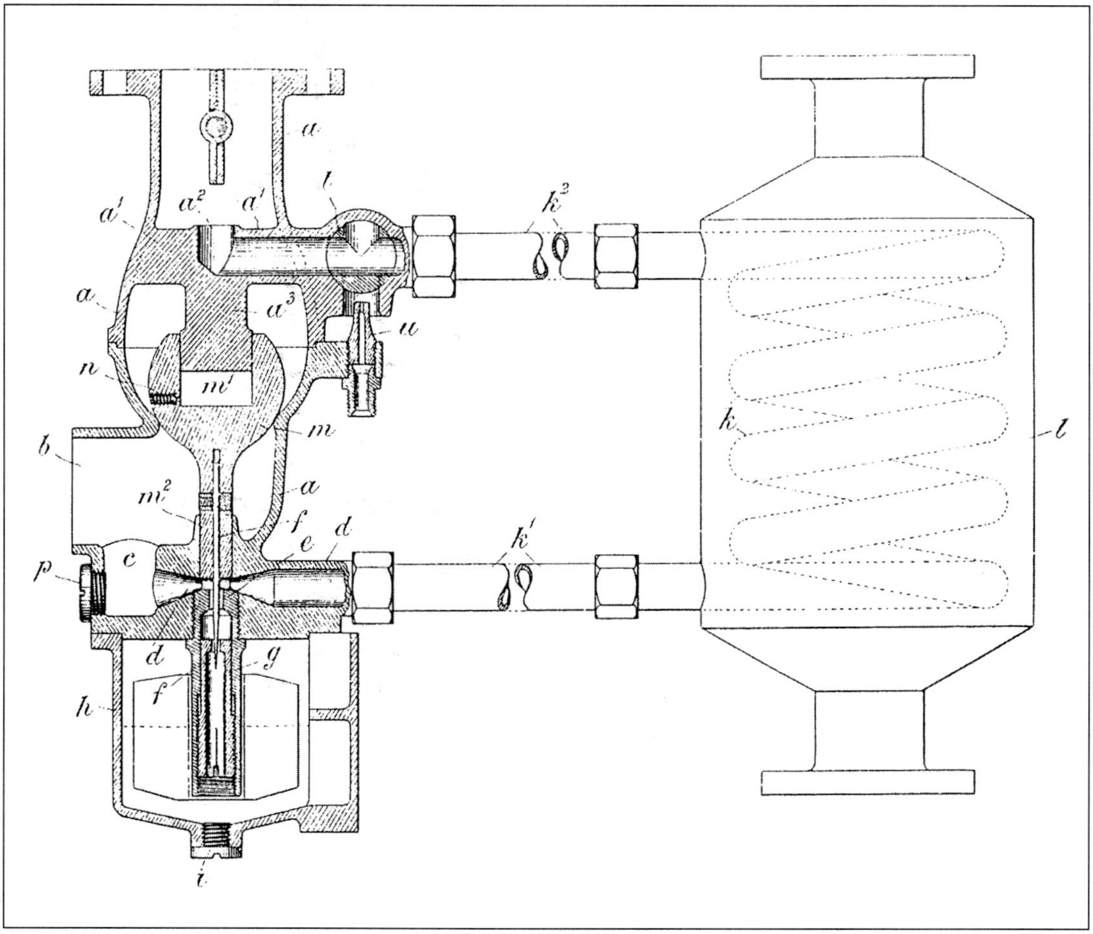

Fig. 1

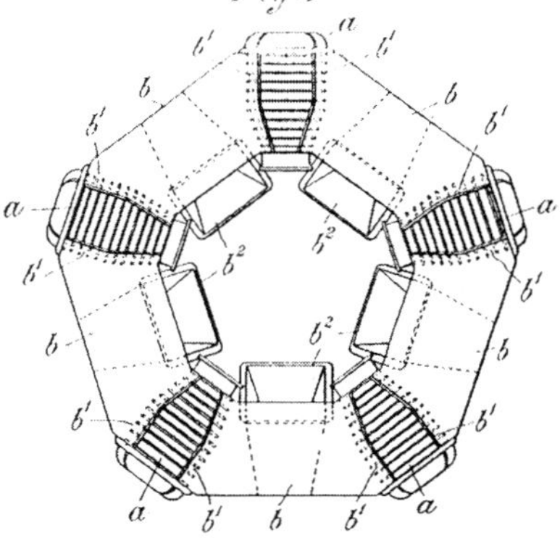

Fig. 3.

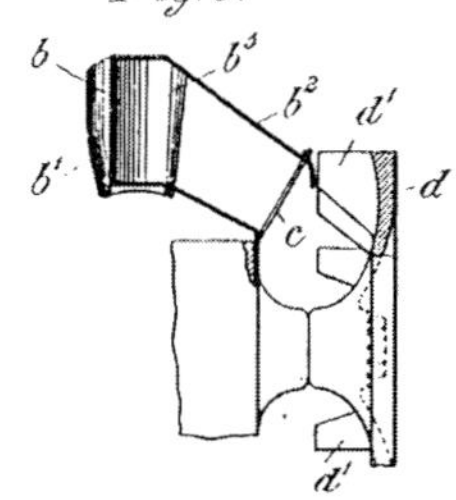

Fig. 2.

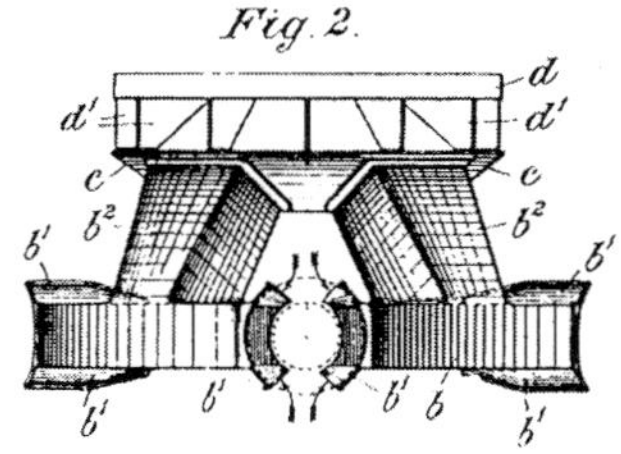

Fig. 4.

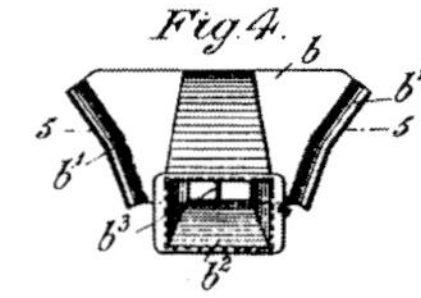

Fig. 5.

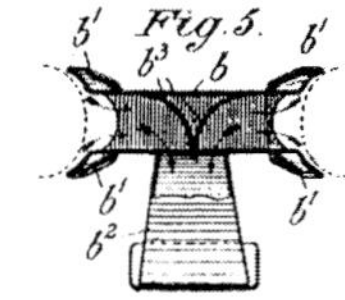

Fig. 1

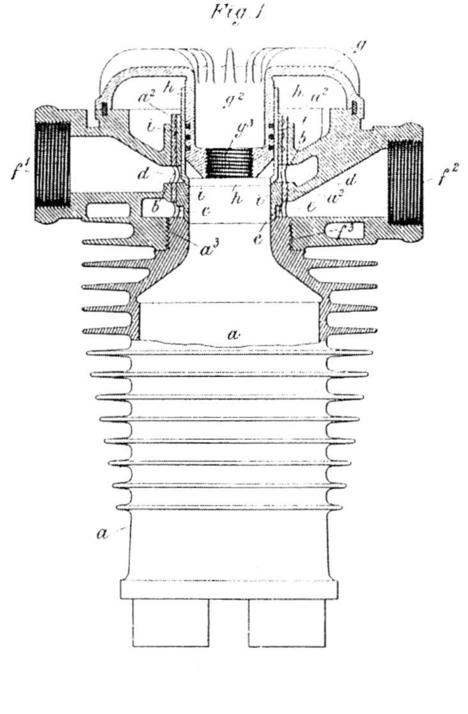

Fig. 3.

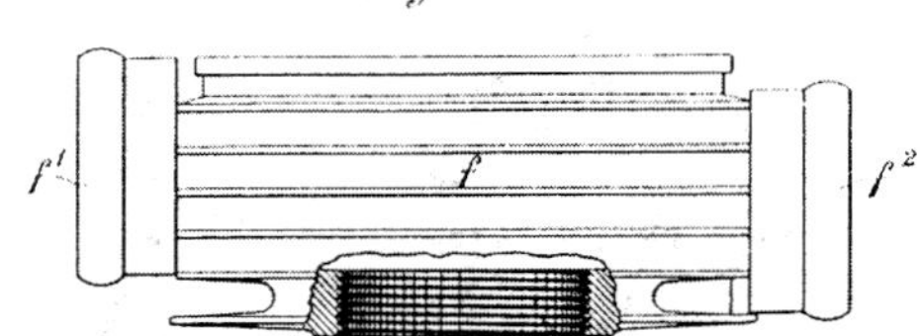

Fig. 2.

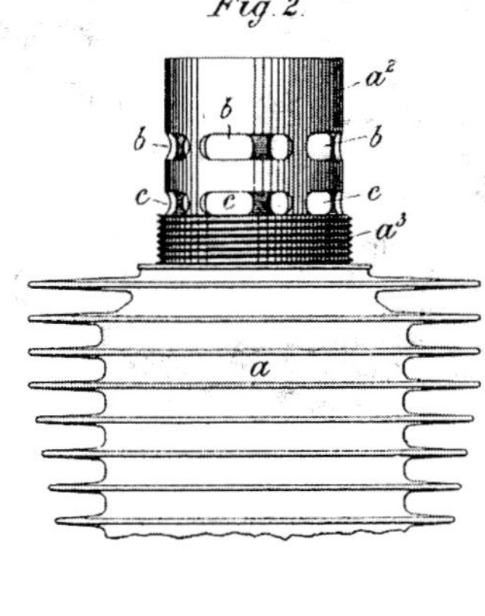

LEFT-TOP: PATENT NUMBER 137439
Taken out jointly in March 1919 by Arthur William Reeves - Manager, and Charles Edward Simms - Mechanical Engineer & Designer, both of the Matchless Works, for the design of air cooling arrangements associated specifically with a 5 cylinder air cooled motor car engine of radial configuration.

LEFT-BOTTOM: PATENT NUMBER 154423.
Taken out in November 1919 by Arthur William Reeves, to cover the design of an air cooled engine cylinder and cylinder head.

BELOW: PATENT NUMBER 158371.
Taken out in November 1919 by Arthur William Reeves and Charles Edward Simms, to cover the design of mechanically operated sleeve valves in an air cooled engine cylinder head. All three of these patents were directly associated with the ill-fated 10 h.p. radial engined motor car which Alldays were to introduce shortly after the First World War, a motor car which very nearly bankrupted the company.

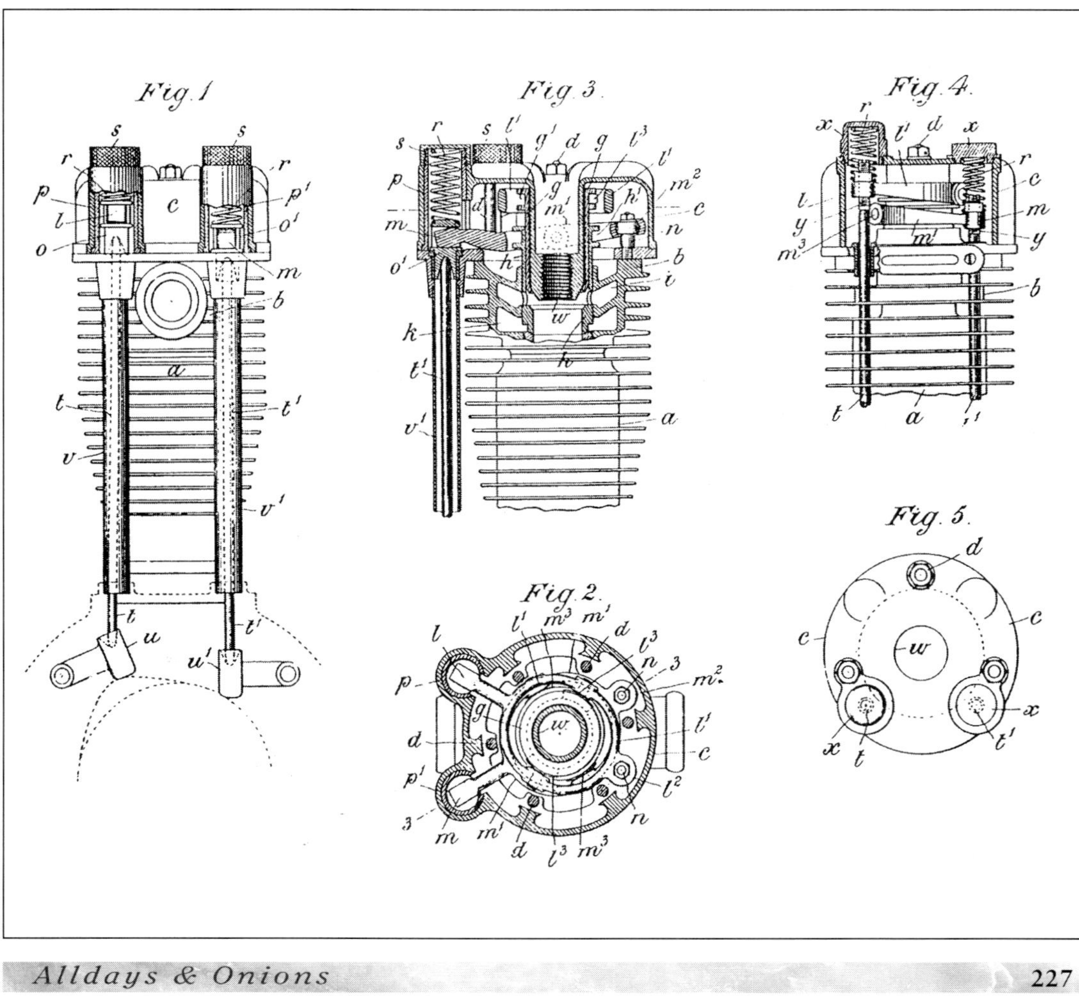

APPENDIX

This Appendix contains photographs and/or information which, whilst not directly related to the Alldays & Onions factories in the time span covered by the general text, is recorded here for the interest of readers.

This section also contains photographs of the Great Western Works and Waverley Works taken in 1998.

PHOTOGRAPHS

LEFT-TOP: A picture taken in 1998 looking up Sydenham Road towards the junction of Anderton Road and Montgomery Street, with the Great Western Works on the right, now barely recognizable with a flat roof on the main office buildings and the upper floor removed completely from the section of the building to the far right. These major alterations to the original buildings took place during, or just after, the Second World War. The gable ended building just visible to the extreme left of this picture is the "Great Western Hall" built in 1896 as a Mess Room and, from the outside at least, little altered in its overall appearance.

LEFT-BOTTOM: Also taken in 1998, this row of shops, running from the corner of Sydenham Road along Anderton Road, is all that remains of the houses Alldays built for their employees in the first phase of this development in 1894. The conversion to shops is believed to have taken place between the wars and was carried out by simply building large bay windows on the front of each property.

ILLUSTRATION

BELOW: Drawing from Patent Number 16822 taken out jointly by Robert Henry Davis, a submarine engineer, and Edmund Allday, in 1913. The patent covers the design of an improved apparatus for artificial respiration using two small bellows and an oxygen bottle to remove toxic gases from a patient's lungs and supply oxygen in its place.

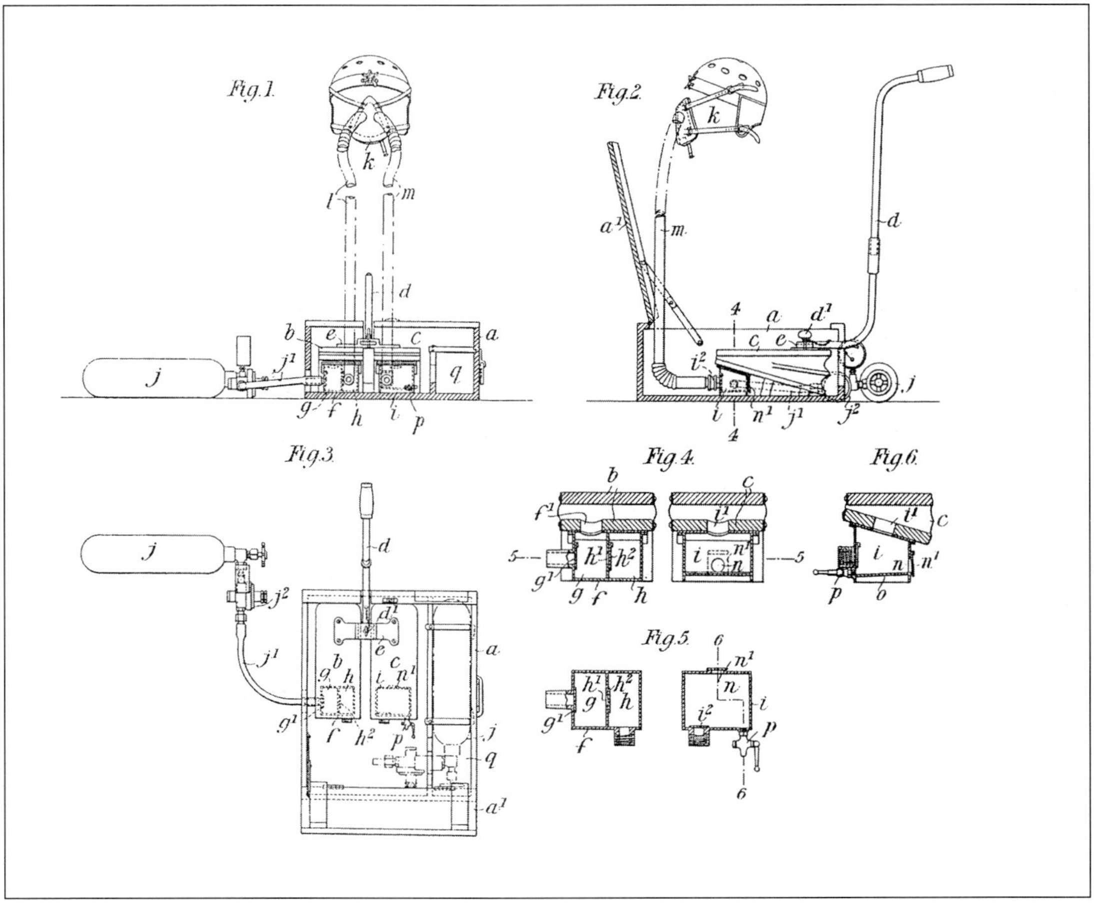

LEFT-TOP: Taken in 1998 from the canal side this 3 storey building was built in 1912 in two 25 feet wide bays and stands 54 feet high. The canal is little used nowadays, but one can imagine that it was a hive of activity years ago with barges tying up at the wharf alongside this building delivering coal, sand and iron ore, and taking away the finished products from the works.

LEFT-BOTTOM: The first floor of this building, which was originally an iron plate shop, now stands desolate.

BELOW: The top floor of this same building, originally used as stores and also for work on bellows, with some of the old overhead lineshafts and pulleys still mounted high in the roof space as a reminder of past activities in this part of the old works which is home now for little more than a few pigeons!

LEFT-TOP: The ground floor of the 3 storey building still has the two overhead electric cranes in situ used for lifting heavy components about the workshop when it was in use as a heavy machine shop.

LEFT-BOTTOM: Although now covered by a layer of modern bricks this is the only surviving building of the Waverley Works, situated just off the Small Heath bypass. Built in 1914 and measuring 240 feet by 200 feet this was the body building and mounting shop where all the motor car and commercial vehicle bodies were assembled and mounted onto chassis. Now converted to small factory units it was somewhat ironic to find that a small sheet metal fabricating business in this building had supplied parts to the Alldays Peacock & Co.Ltd. works in Sydenham Road and Weston-super-Mare.

BELOW: The front of the "Great Western Hall" building still standing opposite what remains of the old Sydenham Road Great Western Works. The entrance to the left accessed the ground floor and shop which is situated between the two entrances, and the entrance to the right accessed stairs to the upper floor. The building was designed by Ewen Harper of Colmore Row in Birmingham and a descendant of this company still practises as an architect in the city.

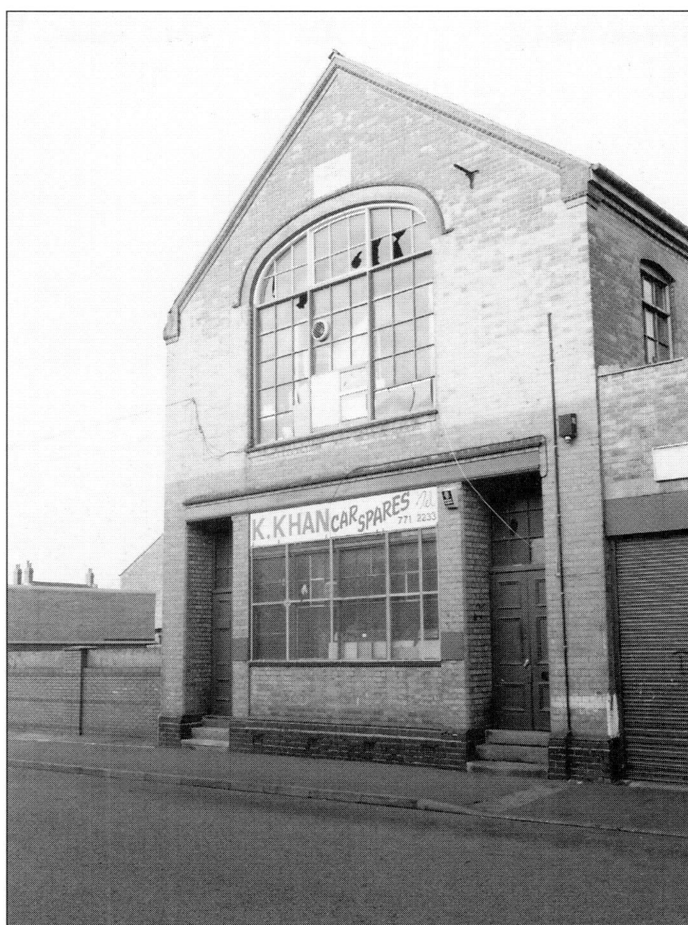

ILLUSTRATION

LEFT: A contemporary advertisement for William Allday & Co. Ltd., the company started by Percy William Allday after resigning from Alldays & Onions in 1921. It is believed that he started up originally as an engineering consultant, but soon decided to set up a new manufacturing business, producing a similar range of engineering products to that of his old family business.

PHOTOGRAPH

BELOW: Perhaps the most pleasant way to conclude this history of one of the oldest engineering companies in Great Britain is to reproduce this photograph of Michael and Mary Allday sitting in their 1903 $7^1/_2$ h.p. single cylinder Alldays & Onions motor car.

LANDMARK COLLECTOR'S LIBRARY

LANDMARK
Publishing Ltd

Ashbourne Hall, Cokayne Ave, Ashbourne, Derbyshire, DE6 1EJ England
Tel 01335 347349 Fax 01335 347303
e-mail landmark@clara.net web site: www.landmarkpublishing.co.uk

Mining Histories

- Collieries of South Wales: Vol 1 *ISBN: 1 84306 015 9, £22.50*
- Collieries of South Wales: Vol 2 *ISBN: 1 84306 017 5, £19.95*
- Collieries of Somerset & Bristol *ISBN: 1 84306 029 9, £14.95*
- Copper & Lead Mines around the Manifold Valley, North Staffordshire *ISBN: 1 901522 77 6, £19.95*
- Images of Cornish Tin *ISBN: 1 84306 020 5, £29.95*
- Lathkill Dale, Derbyshire, its Mines and Miners *ISBN: 1 901522 80 6, £8.00*
- Rocks & Scenery the Peak District *ISBN: 1 84306 026 4, paperback, £7.95*
- Swaledale, its Mines & Smelt Mills *ISBN: 1 84306 018 3, £19.95*

Industrial Histories

- Alldays and Onions *ISBN: 1 84306 047 7, £24.95*
- The Life & Inventions of Richard Roberts, 1789 -1864 *ISBN: 1 84306 027 2, £29.95*
- The Textile Mill Engine *ISBN: 1 901522 43 1, paperback, £22.50*
- Watt, James, His Life in Scotland, 1736-74 *ISBN 1 84306 045 0, £29.95*
- Wolseley, The Real, Adderley Park Works, 1901-1926 *ISBN 1 84306 052 3, £19.95*
- Morris Commercial *ISBN: 1 84306 069 8 (Price to be announced)*

Roads & Transportantion

- Packmen, Carriers & Packhorse Roads *ISBN: 1 84306 016 7, £19.95*
- Roads & Trackways of Wales *ISBN: 1 84306 019 1, £22.50*
- Welsh Cattle Drovers *ISBN: 1 84306 021 3, £22.50*
- Peakland Roads & Trackways *ISBN: 1 901522 91 1, £19.95*

Regional/Local Histories

- Colwyn Bay, Its History across the Years *ISBN: 1 84306 014 0, £24.95*
- Crosses of the Peak District *ISBN 1 84306 044 2, £14.95*
- Derbyshire Country Houses: Vol 1 *ISBN: 1 84306 007 8, £19.95*
- Derbyshire Country Houses: Vol 2 *ISBN: 1 84306 041 8, £19.95*
- Historic Hallamshire *ISBN: 1 84306 049 3, £19.95*
- Llandudno: Queen of Welsh Resorts *ISBN 1 84306 048 5, £15.95*
- Llanrwst: the History of a Market Town *ISBN 1 84306 070 1, £14.95*
- Lost Houses in and around Wrexham *ISBN 1 84306 057 4, £16.95*
- Lost Houses of Derbyshire *ISBN: 1 84306 064 7, £19.95, October 02*
- Shipwrecks of North Wales *ISBN: 1 84306 005 1, £19.95*
- Shrovetide Football and the Ashbourne Game *ISBN: 1 84306 063 9, £19.95*
- Well Dressing *ISBN: 1 84306 042 6, Full colour, £19.95*